The Biggest Stick: The Employment of Artillery Units in Counterinsurgency

A thesis presented to the Faculty of the U.S. Army
Command and General Staff College in partial
fulfillment of the requirements for the
degree

MASTER OF MILITARY ART AND SCIENCE
Art of War

By

RICHARD B. JOHNSON, MAJOR, UNITED STATES ARMY
M.A., Webster University, Saint Louis, Missouri, 2006
B.S., United States Military Academy, West Point, New York, 1999

Fort Leavenworth, Kansas
2011-01

The cover photo courtesy of the Library of Congress is that of General Dwight Eisenhower giving orders to American paratroopers in England.

Abstract
The Biggest Stick:
The Employment of Artillery Units in Counterinsurgency
By Major Richard B. Johnson

This study uses a comparative analysis of the Malayan Emergency, the American experience in Vietnam, and Operation IRAQI FREEDOM to examine the role and effectiveness of artillery units in complex counterinsurgency environments. Through this analysis, four factors emerge which impact the employment of artillery units: the counterinsurgency effort's requirement for indirect fires; constraints and limitations on indirect fires; the counterinsurgency effort's force organization; and the conversion cost of nonstandard roles for artillery units. In conclusion, the study offers five broadly descriptive fundamentals for employing artillery units in a counterinsurgency environment: invest in tactical leadership, exploit lessons learned, support the operational approach and strategic framework, maintain a pragmatic fire support capability, and minimize collateral damage. Finally, the study examines the role of education for leaders in a counterinsurgency, and its influence on these imperative fundamentals.

Objectives of the Art of War Scholars Program

The Art of War Scholars Program is a laboratory for critical thinking. It offers a select group of students a range of accelerated, academically rigorous graduate level courses that promote analysis, stimulate the desire for life-long learning, and reinforce academic research skills. Art of War graduates will not be satisfied with facile arguments; they understand the complexities inherent in almost any endeavor and develop the tools and fortitude to confront such complexities, analyze challenges, and independently seek nuanced solutions in the face of those who would opt for cruder alternatives. Through the pursuit of these outcomes, the Art of War Scholars Program seeks to improve and deepen professional military education.

The Art of War Program places contemporary operations (such as those in Iraq and Afghanistan) in a historical framework by examining earlier military campaigns. Case studies and readings have been selected to show the consistent level of complexity posed by military campaigns throughout the modern era. Coursework emphasizes the importance of understanding previous engagements in order to formulate policy and doctrinal response to current and future campaigns.

One unintended consequence of military history education is the phenomenon of commanders and policy makers "cherry picking" history—that is, pointing to isolated examples from past campaigns to bolster a particular position in a debate, without a comprehensive understanding of the context in which such incidents occurred. This trend of oversimplification leaves many historians wary of introducing these topics into broader, more general discussion. The Art of War program seeks to avoid this pitfall by a thorough examination of context. As one former student stated: "The insights gained have left me with more questions than answers but have increased my ability to understand greater complexities of war rather than the rhetorical narrative that accompanies cursory study of any topic."

Professor Michael Howard, writing "The Use and Abuse of Military History" in 1961, proposed a framework for educating military officers in the art of war that remains unmatched in its clarity, simplicity, and totality. The Art of War program endeavors to model his plan:

Three general rules of study must therefore be borne in mind by the officer who studies military history as a guide to his profession and who wishes to avoid pitfalls. First, he must study in **width**. He must observe the way in which warfare has developed over a long historical period. Only by seeing what does change can one deduce what does not; and as much as can be learnt from the great discontinuities of military history as from the apparent similarities of the techniques employed by the great captains through the ages....Next he must study in **depth**. He should take a single campaign and explore it thoroughly, not simply from official histories, but from memoirs, letters, diaries. . . until the tidy outlines dissolve and he catches a glimpse of the confusion and horror of real experience... and, lastly, he must study in **context**. Campaigns and battles are not like games of chess or football matches, conducted in total detachment from their environment according to strictly defined rules. Wars are not tactical exercises writ large. They are... conflicts of societies, and they can be fully understood only if one understands the nature of the society fighting them. The roots of victory and defeat often have to be sought far from the battlefield, in political, social, and economic factors which explain why armies are constituted as they are, and why their leaders conduct them in the way they do....

It must not be forgotten that the true use of history, military or civil... is not to make men clever for the next time; it is to make them wise forever.

Gordon B. Davis, Jr.
Brigadier General, US Army
Deputy Commanding General
CAC LD&E

Daniel Marston
DPhil (Oxon) FRHistS
Ike Skelton Distinguished Chair
in the Art of War
US Army Command & General
Staff College

Acknowledgments

The largest amount of gratitude possible goes to my ladies: Lea, Lucia, and Stella. Unfortunately you shared my attention with the large volumes of writing listed in the Bibliography. The last class told us at the beginning that "it is only a lot of reading if you do it." With your grace, I did it. Without your support and love, this paper would not have been completed, much less started.

To Aaron, Ben, Dustin, Mark, Nate, Rob, and Tom. Thanks for your patience as I tried to explain my incoherent themes while this project was in its infancy. Appreciation is also in order for our brothers in arms in both the British and American Armies, who extended us the best in hospitality, camaraderie, and discussion during our research trips.

Thanks to the real experts who helped: Bruce Gudmundsson at Quantico for taking time out to discuss all things artillery, John Dubuisson in the CARL for showing me how to get to "the goods" in military history archives, and especially Dr. Daniel Marston for making sense of it all when I could not.

Finally, to the Paratroopers of the 1st Battalion, 319th Airborne Field Artillery Regiment and all of the attachments during the various iterations of Task Force Loyalty. Counterinsurgency is equal parts combating insurgents and combating a longing for home; I am in unwavering debt to you for the efforts with both. This work is dedicated to your sacrifices during *The Long War*. I have looked through a good portion of history and I cannot find another artillery unit that embodies the character of adaptiveness, perseverance, and determination in counterinsurgency better than this group...although I may be a bit biased.

Contents

Abstract iii
Acknowledgments vi
Table of Contents vii
Glossary viii

Chapter 1 Introduction 1

Chapter 2: The Themes of Counterinsurgency and Artillery Units 7

Chapter 3: The Malayan Emergency 39

Chapter 4: The American Experience in Vietnam 65

Chapter 5: Operation IRAQI FREEDOM 117

Chapter 6: Conclusions 187

Bibliography 197

Glossary

AAB	Advise and Assist Brigade
AEI	American Enterprise Institute
ACR	Armored Cavalry Regiment
APC	Accelerated Pacification Program
AQI	*Al Qaeda* in Iraq
ARA	Aerial Rocket Artillery
ARVN	Army of the Republic of Vietnam
BCT	Brigade Combat Team
CAP	Combined Action Platoon
CDE	Collateral Damage Estimate
CEA	Captured Enemy Ammunition
CENTCOM	Central Command
CIA	Central Intelligence Agency
CIDG	Civilian Irregular Defense Group
CJTF	Coalition / Joint Task Force
CORDS	Civil Operations and Revolutionary Development Support
COSVN	Central Office for South Vietnam
CPA	Coalition Provisional Authority
CT	Communist Terrorist
CTZ	Corps Tactical Zone
DMZ	De-Militarized Zone
DWEC	District War Executive Council
EFP	Explosively Formed Projectile
FA	Field Artillery
FOB	Forward Operating Base
FSB	Fire Support Base
GMLRS	Guided Multiple Launch Rocket System
GOC	General Officer Commanding
GoI	Government of Iraq
H&I	Harassment and Interdiction
HUMINT	Human Intelligence
IA	Iraqi Army
IED	Improvised Explosive Device

ING	Iraqi National Guard
ISF	Iraqi Security Force(s)
JAM	*Jaysh al Mehdi*
JCS	Joint Chiefs of Staff
JSS	Joint Security Station
MAAG	Military Advisory and Assistance Group
MACV	Military Assistance Command, Vietnam
MCP	Malayan Communist Party
MiTT	Military Training Team or Military Transition Team
MLRS	Multiple Launch Rocket System
MNC-I	Multi-National Corps - Iraq
MND-B	Multi-National Division - Baghdad
MNF-I	Multi-National Force - Iraq
MNSTC-I	Multi-National Security Transition Command - Iraq
MPABA	Malayan Peoples' Anti-British Army
MRLA	Malayan Races Liberation Army
NP	National Police
NVA	North Vietnamese Army
OCO	Office of Civil Operations
OIF	Operation Iraqi Freedom
OND	Operation New Dawn
ORHA	Office for Reconstruction and Humanitarian Assistance
PF	Popular Forces
PGM	Precision Guided Munition
PRT	Provincial Reconstruction Team
RA	Royal Artillery
RF	Regional Forces
ROE	Rules of Engagement
RPG	Rocket-Propelled Grenade
SAS	Special Air Service
SEP	Surrendered Enemy Personnel
SOI	Sons of Iraq
SVN	South Vietnam
SWEC	State War Executive Council

USAID	United States Agency for International Development
USMC	United States Marine Corps
VBIED	Vehicle Borne Improvised Explosive Device
VC	Vietcong
WMD	Weapon(s) of Mass Destruction

ARTILLERY UNIT: A brigade-, battalion-, or battery-level force capable of providing lethal indirect fires by means of integrated fire support tasked to maneuver units; a fire direction element; and assigned guns, howitzers, or launchers to deliver lethal fires.

COUNTERINSURGENCY: The actions taken by a government to limit or defeat an insurgency.

FIRE SUPPORT: The fires that support land, maritime, amphibious, and special operations forces to engage enemy forces, combat formations, and facilities in pursuit of tactical and operational objectives.

FIRES: The effects of lethal or nonlethal weapons.

INDIRECT FIRES: Fire delivered on a target that is not itself used as a point of aim for the weapon or the director.

INSURGENCY: An organized movement aimed at the overthrow of a constituted government through the use of subversion and armed conflict.

OPERATIONAL APPROACH: The use of operational art in a specific environmental context; the pursuit of a strategic objective by arranging tactical actions in time, space, and purpose.

Chapter 1
Introduction

> Modern wars are not internecine wars in which the killing of the enemy is the object. The destruction of the enemy in modern war, and, indeed modern war itself, are means to obtain that object of the belligerent which lies beyond the war.
>
> — Deputy Chief of Staff for Military Operations, *Program for the Pacification and Long-Term Development of South Vietnam*

> I have noticed that the mere mention of the word "pacification" to a group of soldiers, whatever their rank, usually brings forth deriding smiles. Many of them seem to think of pacification as the distribution of candies to the children and smiles to the old people. We certainly must show the carrot in our left hand, but only if we brandish a stick in our right hand. If skepticism about pacification is prevalent today, it is due to the fact that the stick has been too inconspicuous until now, or used too haphazardly and without a plan.
>
> — David Galula, *Pacification in Algeria*

Artillery, The Counterinsurgent's Biggest Stick

One aspect of successful counterinsurgency efforts is the amalgamation of attractive "carrots" and coercive "sticks," even if these sticks are not necessarily lethal combat power.[1] Artillery units provide a counterinsurgency effort with the ability to brandish the stick of massed indirect fires. Since the advent of modern firepower, it has been a key element in warfare as practiced by western cultures.[2] In counterinsurgency warfare, there are few sticks larger than the ability to leverage accurate and predicted indirect lethal fires on an insurgent force among the population. Conversely, there are also few responsibilities higher than the requirement to minimize civilian suffering as a by-product of lethal action. To a certain degree, this is a reflection of the counterinsurgent's imperative to sensibly restore the societal monopoly on violence to the governing power. But refined counterinsurgency approaches are not about the presence of attractive and coercive means, they are about the manner of employing those means with a nuanced understanding of their effects. As such, the employment of artillery units warrants a detailed analysis, especially in an era when the guerrilla and the physicist seemed to conspire to push "normal" warfare into the dustbin of history.[3]

Modern artillery is at a crossroads, but not a crisis. Senior leaders identify both the need to regain the core competencies of indirect fire proficiency after years of service in nonstandard roles, and the need to integrate this institutional experience in other missions.[4] No analyst or strategist can faithfully predict the next war with complete confidence, so the need remains for flexible forces that are rooted in their primary combat functions. This requisite flexibility is found in tactical leaders who are broadly educated, to confidently put their experience and training into context in an amorphous and uncertain environment. It is quite possible that in the next conflict, it will not be the side with the best technology, training, or information that achieves their strategic objectives; it may be the side with the most competent leaders.[5]

Factors and Fundamentals

This study uses comparative analysis across three case studies to identify descriptive themes of the employment of artillery units in counterinsurgencies. This is not an attempt to isolate the prescriptive principles which govern the most efficient means of their utilization in a specific, defined form of warfare. By identifying trends while accounting for the peculiarities in each environment, four factors emerge which influence the integration of artillery units:

1. The counterinsurgency effort's requirement for indirect fires.

2. Constraints and limitations on indirect fires.

3. The counterinsurgency effort's force organization.

4. The conversion cost of nonstandard roles for artillery units.

These factors are developed further in Chapter Two and form the framework to analyze their employment in each case study.

Historical case studies provide an effective vehicle for comparative analysis. In the Malayan Emergency, British artillery units showed that a counterinsurgent force can achieve lethal effects through a practical, limited, and decentralized employment. The American experience in Vietnam illustrates that even the most capable and adaptive massed fires do not address the drivers of instability in an insurgency if they are part of an ill-suited operational approach and strategic framework. The third case study, Operation IRAQI FREEDOM, examines the practical use of artillery units in many different roles, the organizational conversion cost of switching mission profiles, and the long-term effects on the units themselves.

The cumulative analysis of these case studies yields five broadly descriptive fundamentals which illustrate the imperatives for implementing artillery units in a counterinsurgency effort. These fundamentals include the requirements to:

1. Invest in artillery units' tactical leadership.

2. Exploit lessons learned.

3. Support the operational approach and strategic framework.

4. Maintain a pragmatic fire support capability.

5. Minimize collateral damage.

These five fundamentals represent actions in a counterinsurgency, but they also describe actions which must pervade training and preparations prior to a counterinsurgency operation in order to be truly effective.

This study is limited by a reliance on primary sources to provide sufficient accuracy and credibility.[6] The Malayan Emergency, the American experience in Vietnam, and Operation IRAQI FREEDOM (OIF) provide an appropriate level of variation and temporal separation to make meaningful comparisons. The first step in analyzing these counterinsurgencies is to examine the themes of counterinsurgency and artillery to establish a consistent lexicon and framework for analysis.

Notes

1. BI050, Dhofar Veterans Panel, Interview by Mark Battjes, Ben Boardman, Robert Green, Richard Johnson, Aaron Kaufman, Dustin Mitchell, Nathan Springer, and Thomas Walton, United Kingdom, 28 March 2011. An SAS veteran of the Dhofar campaign alternatively described this dichotomy as "the velvet glove and the mailed fist," which is a closer metaphor since it describes two symmetric elements (hands) with different equipment and inherent training (glove and mail). This study uses the carrot and stick analogy due to its relative familiarity and ubiquitous presence in counterinsurgency literature.

2. Robert Scales' preface to *Firepower in Limited War* discusses this phenomenon at length. Refer to Robert H. Scales, *Firepower in Limited War* (Novato, CA: Presidio Press, 1995), ix-xi. For an illustration of the dominance of technology and discipline in the "western way of war," see Geoffrey Parker, "The Western Way of War," in *The Cambridge History of Warfare*, ed. Geoffrey Parker (New York: Cambridge University Press, 2005), 1-3. For a broader discussion on the relevance of culture in approaches to warfare, see Patrick Porter, *Military Orientalism: Eastern War Through Western Eyes* (New York: Columbia University Press, 2009), 2, 15-16, 193-195.

3. Bruce I. Gudmundsson, *On Artillery* (Westport, CT: Praeger, 1993), 153. This quote is paraphrased from Gudmundsson's description of the environment in Vietnam, reflecting the presence of hyper-organized Communist insurgents and the specter of nuclear war: "In an era when the guerrilla and the physicist seemed to conspire to push 'normal' warfare into the dustbin of history." The description holds true to this day, and begs the reader to contemplate if "normal war" even exists any more with the emergence of increased irregular (and hybrid) warfare, weapons of mass destruction, and global connectedness.

4. Michael S. Tucker and Jason P. Conroy, "Maintaining the Combat Edge," *Military Review* 91, No. 3 (May-June 2011): 8-15; Patricia Slayden Hollis, "Division Operations Across the Spectrum-Combat to SOSO in Iraq: Interview with Major General Raymond T. Odierno, CG of 4th ID in OIF," *Field Artillery* (March-June 2004): 11. In their discussion on the effort to recapture atrophied skills for conventional warfare, Major General Tucker and Major Conroy quote the Secretary of Defense, Robert M. Gates: "The army has to regain its edge in fighting conventional wars while retaining what it has learned about fighting unconventional wars." In General Odierno's 2004 interview with *Field Artillery*, he asserted that "artillery has to be a versatile asset. The Army can no longer afford to have artillerymen just do artillery missions. So Redlegs also must be able to set up flash checkpoints, patrol, conduct cordon and search operations, etc."

5. Mark Moyar advances this theory in his book *A Question of Command*, in which he describes a leader-centric approach to counterinsurgency warfare, with the attributes of effective counterinsurgent leaders being initiative, flexibility, creativity, judgment, empathy, charisma, sociability, dedication, integrity, and organization. Also, refer to Mark Moyar, "Development in Afghanistan's Counterinsurgency," *Small Wars Journal*, www.smallwarsjournal.com/blog/2011/03/development-in-afghanistans-co/ (accessed 21 March 2011).

6. A complete investigation into the employment of artillery units in counterinsurgencies should include the Soviets in Afghanistan, the French in Algeria,

and the Israeli Defense Force operations in Gaza and Lebanon, but the unavailability of sufficient resources in English prevents their inclusion in this study. Additionally, the record is incomplete without a discussion on the Coalition effort in Afghanistan (Operation ENDURING FREEDOM) which should be conducted once artillery units' operations there are complete and there is sufficient standoff to conduct an analysis.

Chapter 2
The Themes of Counterinsurgency and Artillery Units

> All eyes turn to the soldier when violence erupts. Before this happens warnings given by the army and precautions recommended, are sometimes ignored or treated with disdain. It is the course of human nature to avoid the unpleasant for as long as possible, and potential insurgents are sure to give the public every encouragement to stick their heads in the sand whilst their preparations are in progress. But when the fighting starts the soldier will not only be expected to know how to conduct operations, he may also have to advise on government measures as well.
>
> — General Sir Frank Kitson, *Bunch of Five*

> The more important the subject and the closer it cuts to the bone of our hope and needs, the more we are likely to err in establishing a framework for analysis.
>
> — Steven J. Gould, *Full House*

This chapter discusses the nature of counterinsurgency and artillery units. To provide clarity and a realistic scope for this research, several definitions and distinctions are required regarding the elements of counterinsurgency. This chapter examines counterinsurgency within the spectrum of warfare, in order to inform the review of existing theory and contemporary issues. These themes culminate with the four factors which influence the employment of artillery units in counterinsurgencies, to provide a framework for consistent analysis in further case study comparisons.

Counterinsurgency as War

Although some of the component activities of insurgency and counterinsurgency are nonmilitary in nature, counterinsurgency is still a method of prosecuting conflict within the spectrum of warfare. It is not separate from war, nor is it a complementary approach to war. The inclusion of social and political aspects in counterinsurgency does not remove it from the spectrum of warfare since war itself is a means to social and political ends. Several of the most influential military theorists in history cast insurgency and counterinsurgency within the larger subject of war, namely Clausewitz, Jomini, and Sun Tzu.

Carl von Clausewitz[1] was a Prussian military officer and theorist whose writings have gained steady momentum within military academia over the past century. Now considered one of the classics on military thought, his work *On War* contains several invaluable perspectives on the nature of warfare itself. Although most of Clausewitz's tactical theories are only marginally applicable today, his combination of intellectual realism and aestheticism with respect to the nature and strategies of war remain instructive to modern political and military thinkers. His contemporary paradigm of Westphalian nation-states colors his discussions on military strategy, but not his discussion on the nature of warfare itself. His work retains relevancy even though it is now separated from the present by the industrial revolution, two global conflicts and the collapse of a bipolar world order.[2]

For critics who see counterinsurgency as a chiefly political endeavor with limited military support, it is crucial to understand that war itself is political. Clausewitz successfully demonstrates that war should be the basic struggle for a political objective. This is true in the context of a large conventional state-on-state war and in a localized insurgency. Instead of the pursuit of vital national requirements, the political objective in counterinsurgency is "the weakening or displacing of a legitimate government," as this study defines it. Clausewitz contends that when people go to war, their cause is political. He concludes that "war, therefore, is an act of policy."[3] In this framework of war as an act of policy, Clausewitz continues to characterize the conduct of war as a continuation of an ongoing political struggle via alternate means.[4] This is reflected in his definition of war, as he states that war is "an act of force to compel our enemy to do our will."[5]

Separate from his descriptions of the political nature of war, Clausewitz shows an understanding for the human aspect of counterinsurgency since it is waged amongst the population. He also acknowledges that the population is a critical resource, much more than just a recruiting pool for either side of the conflict. He discusses the population itself as a third of his paradoxical trinity expressed by the people, the military and the dominant government policy. Clausewitz concludes that "the passions that are kindled in war must already be inherent in the people."[6] Although *On War* does not focus on irregular warfare as we would recognize it today, his earlier lectures on small wars (the *Kleinkrieg*) and guerilla warfare (the *Volkskreig*) are integrated into his short passage "The People in Arms."[7]

Where Clausewitz's focus is on the broadly descriptive elements of strategy, Swiss military theorist Baron Antoine de Jomini[8] provides a somewhat surprising perspective that casts counterinsurgency as a broad form of war; surprising because his works reflect the prescriptive nature of a scientific reductionist. Jomini was Clausewitz's contemporary and generally on the winning side of similar campaigns, and as such he was more engaged in the practice of distilling the keys to victory into tactical principles. However, Jomini's own experience as an "eager revolutionary" in the Swiss revolution of 1798 would have been extremely influential, as this is when he began studying military art.[9] This is reflected in his discussions regarding insurgency and counterinsurgency, which he considers together as "national wars," or alternatively "civil and religious wars."[10]

Jomini recognizes the uniqueness of wars of insurgency and counterinsurgency amongst the population, but does not cast them in a different class: "Two hundred thousand French wishing to subjugate Spain, aroused against them as one man, would not maneuver like two hundred thousand French wishing to march upon Vienna."[11] He identifies the necessity to alter military tactics, but he does not insist that this makes it an activity separate from war. In his characterization of these national wars, Jomini continues to show the difficulties that face a counterinsurgency effort when such an aroused population is backed by a core of disciplined troops.[12] Although he lives up to his main criticism of reducing warfare to a set of rules, Jomini shows an appreciation for counterinsurgency as an act of war, not strictly a social and political activity.[13]

Another influential military theorist that focuses on prescriptive tactical elements while understanding the nature of counterinsurgency is the Chinese military theorist Sun Tzu.[14] Although he writes instructive principles for a field general, his work is based on the human element of warfare. This is a large part of what makes his theories applicable today, since human nature does not change with the rapidity of technology and tactics.[15] Although Sun Tzu does not directly address revolutions or insurgencies, his principles show a deep understanding of war beyond the physical battlefield itself. Since Sun Tzu focuses his theories in the human dimension and on its participants, this quality makes his theories instructive towards counterinsurgency as well as positional open warfare. Sun Tzu acknowledges the integration of several bases of power which are applicable to successful counterinsurgency. He writes that leaders "must not rely solely on military power, as the fighting on the battlefields is just one front in a total war."[16] Although he wrote over two millennia ago, his principles of attacking weakness, avoiding strength and exercising patience are some of the key tenets of insurgent theory and practice today.[17] Interestingly, they are applicable to the political aspect of counterinsurgency as well as the military aspect.

Insurgency Themes

An endeavor to understand the nature of an uncertain and amorphous activity requires common terms to frame the discussion. There is a delicate balance between the need for precise terminology to accurately convey a theme's limits, and the need for deliberately indefinite terms which ensure the inclusion of many important historical examples. Useful definitions serve as this foundation for analysis, and as such they must have a specific meaning to convey useful information.

Sometimes the need for accuracy in definitions leads to an exclusion of some key components in insurgency and counterinsurgency. As will be discussed later in this chapter, the contemporary debate on population-centric and enemy-centric approaches to counterinsurgency focus on the activity rather than the complimentary components of each approach when considered together. The following definition and themes focus on the relationship between the component parts of each aspect in an attempt to describe the overall activity and organization of insurgency.

Insurgency Defined

Current US Army doctrine defines an insurgency as "an organized movement aimed at the overthrow of a constituted government through the use of subversion and armed conflict."[18] This definition limits insurgencies to those movements that specifically seek to overthrow a government, and unnecessarily restricts the means and objectives that an adversary uses to be considered an insurgent. Therefore, the definition of an insurgency for this study shall be *an organized movement aimed at weakening or displacing a government through any combination of political and armed struggle*. The goal of "displacing" a government reflects the fact that some movements simply seek the redress of a certain grievance or wider autonomy, and not a complete overthrow of a central government. This definition excludes pure social activism and political subversion when conducted alone, since they lack an armed component of the struggle. However, if these activities are pursued with any type of armed component such as terrorism, guerrilla warfare, or open warfare, then the overall effort could be considered a combination of the two forms of action (and therefore an insurgency). Conversely, an armed nonstate component without political considerations is not considered an insurgency. The nature of an insurgency is not linear; it does not necessarily follow a regimented path of disparate tasks which build upon each other sequentially. Defining insurgency simply as an "organized movement" illustrates that these separate actions may be in close concert, or in a distributed network with coincidental goals, provided that there is at least minimal coordination.[19]

Insurgency Theory

Although insurgents have many different grievances or ideological causes, their operations focus on a similar endstate: to weaken or displace a legitimate government. The ideological cause may either avail or obviate some resources specific to that insurgency, but the shared goal of displacing a legitimate government means that insurgency's elements have a degree of portability between theorists. In the search for these elusive elements which support a successful insurgency, many theorists expound on their experiences or observations to unify the themes into a cogent strategy.

Mao Tse-Tung[20] has a primary role among these insurgent theorists. Mao developed a hierarchically organized system to execute an insurgency strategy, writing prodigiously during and after his campaigns against the occupying Japanese forces and Chinese nationalist forces. His works show an astute realization that the underlying cause or ideology has primacy in an insurgency, and that organization along the lines of strict political-military integration was beneficial to his guerrilla effort. This principle extended as far as creating a political officer at the lowest possible echelons to mitigate against the uneven quality of forces raised from the population.[21] One possible criticism of Mao's work is that it is overly prescriptive, since he postulates a detailed structure for a guerrilla force that looks similar to a conventional army's table of organization and equipment.[22]

Mao postulates his theory on the conduct of an insurgency by identifying three phases of warfare. In the first phase, the political movement develops and limited guerilla operations are directly controlled by the political party. The movement's goal is to set conditions among the population and terrain for the following phases of operations, and sees this as a strategically defensive stage in the insurgency.[23] This second phase consists of dispersed guerilla warfare which Mao casts as the strategic stalemate. In this phase, the insurgents focus on establishing secure base areas. One of the guerilla force's goals is to entice the enemy into far-ranging and exhausting search-and-destroy missions. At the close of the second phase, Mao contends that territory can be categorized in three types of area: the enemy base area, the guerilla base area, and the contended area. He sees this protracted second phase as the transitional state in warfare, setting the stage for the "most brilliant last act."[24] Mao's third and final phase marks the transition into conventional warfare (which Mao refers to as "orthodox warfare") against the government force, supported by a continued guerrilla effort. Insurgent forces use a mix of positional and mobile warfare to connect territory to the base area and pressure the government force for capitulation.[25] While guerrilla forces support conventional forces in the third phase, Mao cautions about creating a chasm between the two because he saw them as complementary forces.[26]

Vietnamese General Vo Nguyen Giap[27] refined Mao's theories and adapted them in Vietnam. With external backing from China, Giap structured the military aspects of insurgency in Vietnam with Ho Chi Minh against Japanese occupiers, French colonial forces, and American intervention and support of South Vietnam. Similar to Mao, Giap sees insurgency in terms of a politically-motivated struggle that would eventually build to decisive conventional battles.

One of the clearest consistencies in the Mao and Giap methods is the reliance on the population for all forms of support: manpower, material, information and tacit protection. Giap echoes Mao's analogy that the people play the part of the water where the insurgent forces are fish.[28] Giap sees this in both ideological and pragmatic terms. He shows that an insurgency is the essential form of the people's struggle, and therefore must maintain close contact with their ideology and grievances.[29] In practical terms, he sees the discipline of insurgent forces as the paramount effort to maintain the people's confidence and affection by respecting, helping and defending them.[30]

Giap espouses Mao's three-phase model for insurgency in the same terms, but makes an important distinction regarding the synchronization across an insurgency.[31] Whereas Mao describes the three phases in strictly nationwide strategic terms, Giap describes them in more localized or regional terms.[32] This subtle difference illustrates Giap's ability to implement the phases in different areas in different times. For example, one area may have an emerging political and subversive element with limited guerrilla operations, while a neighboring province could be transitioning from guerrilla operations to conventional warfare. This model reflects Giap's central aim of exhausting the enemy's manpower at its concentrated points without exhausting the limited guerrilla manpower in other areas.[33]

Both Mao and Giap demonstrated the utility of a highly-organized and ideologically motivated insurgency. Interceding insurgencies have not always shared these characteristics, and as such one criticism of these models is their reliance on the organizational and ideological fervor of a Marxist revolution.[34] However, both theorists present their models in terms that have portability to other insurgencies regardless of the source of the conflict. They use their own personal leadership in Marxist civil wars and insurgencies to illustrate the finer points of their insurgency models, but not to define them. After discussing counterinsurgency theory, this chapter examines contemporary debates regarding the changing nature of both insurgencies and counterinsurgencies. Changing trends in ideological causes do not obviate cogent methods of conducting the political and armed struggle in an insurgency.

These insurgent strategies have an important implication for the use of artillery firepower, and by extension the employment of artillery units. Both Mao and Giap identify the population as their base of support and a source of their strength. In a conventional conflict, firepower can target enemy material resources which are considered high payoff targets, but a counterinsurgency cannot target the population writ large in an effort to destroy this base of support. Such efforts would immediately negate any legitimacy. This means that in the first two phases of a Maoist organized counterinsurgency, firepower is generally limited to attacking the insurgent himself and not his critical resources until they establish physical base areas. Consequently, artillery units may have comparatively fewer targets during the longest portion of the campaign, the middle phase in which areas are contested by dispersed guerrilla forces. Since there will inevitably be a much higher demand for indirect fires during the third phase in a counterinsurgency effort against Maoist forces, artillery units could see their roles shifting rapidly.

Contemporary Themes in Insurgency

Insurgency may have much in common with reality television in that both are reflections of a larger society. Accordingly, new theories on insurgency and counterinsurgency rise with changes in society and warfare. An undeniable trend over the past decade is the rise in globally-connected or federated insurgencies.

British theorist John Mackinlay posits that the nodes of a disaffected population that are territorially separated constitute an "Insurgent Archipelago," and that they are now connected by modern communications to become part of the same global movements.[35] In this view, Western powers are unwise to deploy forces in a counterinsurgency intervention before their own populations are secure.[36] This aspect of modern insurgencies does not obviate older theories for counterinsurgency, and Mackinlay acknowledges this by stating that there is a need for a generic threat model to "encourage a uniformly more intensive and sophisticated understanding at the soldier level."[37]

In another attempt to describe the emerging global nature of insurgencies, theorist David C. Gompert advances a model of four types of insurgency. A Type I insurgency is locally or regionally structured, with goals and means existing only within the borders of a given state. A Type II insurgency adds international support, and a Type III insurgency adds coordination with other insurgencies as part of a wider global struggle. Gompert's Type IV insurgencies use these global relationships to target systems of states.[38] Mackinlay's and Gompert's models provide a deep insight to the global nature of insurgencies, but they do not imply that existing counterinsurgency theories must be discarded even if they were authored with a single nation in mind. The notion of unity of effort still applies, perhaps on a larger scale.

Counterinsurgency Themes

The term "counterinsurgency" represents that this effort is more than just an opposing force to insurgency, and more than an "anti-insurgency" aimed at destroying its tangible effects. As the very essence of an insurgency is intangible, a determination to counter it by all means available is a more appropriate theoretical approach. Insurgencies appear in many different manifestations, and as such counterinsurgencies may have many different characteristics to address them.

Counterinsurgency Defined

Keeping in mind that this study's definition of an insurgency is *an organized movement aimed at weakening or displacing a government through any combination of political and armed struggle*, the definition of counterinsurgency for this study will be *the actions taken by a government to limit or defeat an insurgency*. Defining counterinsurgency in these terms allows for a broad range of actions and efforts that governments leverage against insurgents. There is no characterization regarding the supremacy of one source of power over another, since a counterinsurgency effort must consider political, social, economic and military resources among others. With respect to the endstate of a counterinsurgency in this model, allowance is made for a government effort that aims to simply limit the influence of an insurgent group. Specifically, this may prove useful in cases where insurgent groups represent a legitimate grievance within the population, and the government seeks to limit their destructive capability and pursue an approach of accommodation or reconciliation.

This definition deliberately avoids making a distinction of host-nation or intervening counterinsurgent forces from another nation, since these forces may work in concert or apart from each other with the same goals. Additionally, some counterinsurgencies are fought almost entirely by an intervening force on behalf of a militarily weak but legitimate host-nation government. Consider counterinsurgency efforts in Iraq in 2004, when the United States and coalition partners formed the overwhelming majority of military means for security efforts while a sovereign Iraqi government formed slowly.[39]

Counterinsurgency Theory

An overview of counterinsurgency theory will illustrate some common themes in counterinsurgency strategies and operations, and it will provide depth for the case study comparisons in the chapters which follow. The following discussion is only representative of the major theories and elements of counterinsurgency, focused on some of the most enduring and articulate theorists. A complete discussion of counterinsurgency theory would constitute an entirely separate volume of work.

The British Experience and Theoretical Roots

Each nation draws on its unique shared experience to refine strategies for conducting counterinsurgency. This influence is particularly apparent with British counterinsurgency theorists and the unique character of their methods. British counterinsurgency efforts since World War II follow the pattern of earlier colonial interventions where military means were in aid to the civil power. This manifests itself in the theorists' notions that military and political issues must be balanced in a counterinsurgency.[40] British theory also tends toward decentralized leadership and *ad hoc* specialization, in part due to the fact that the British army's regimental system allowed commanders to truly understand their soldiers' strengths and weaknesses after long service together. The British conducted counterinsurgency nearly continuously since 1948, which lends credence that their principles counter more than one specific style of insurgency.[41]

Thompson

Civil administrator and theorist Sir Robert Thompson[42] provides excellent views of counterinsurgent strategy and operations based on his extensive experience in the British Malayan Emergency and in Vietnam during the American intervention. Thompson's theories seek to directly counter a Maoist hyper-organized insurgency, and he views the insurgent struggle as a "war for the people."[43] In this model, he describes the insurgent movement's absolute reliance on a political cause. As the campaign progresses, insurgents may "tack on" many additional minor causes to develop a larger base of support within the population.[44] Thompson makes a delineation between subversion and insurgency in terms of a violent component. He shows that subversion enables an insurgency, but the switch to an armed struggle indicates that peaceful subversion alone is inadequate in most cases.[45] His theories focus on the government-level framework for defeating an insurgency, with a particular focus on organizational models and policy. Thompson advances five principles for counterinsurgency, which apply to insurgencies beyond those with a communist ideology:

1. The government must have a clear political aim: to maintain a stable and viable state.

2. The government must function in accordance with the law.

3. The government must have an overall plan which organizes all efforts.

4. The government must give priority to defeating subversion, not the guerrillas.

5. In the guerrilla phase, the government must secure its base areas first.[46]

Thompson describes counterinsurgency operations in time and space with the sequential phases of clearing, holding, winning, and won. This impacts the employment of artillery units, since they are more instrumental in an indirect fire role during the clearing and holding phases, and more useful in a nonstandard role while the force is winning over the population via expanded governance.[47] Another impact on the employment of artillery units reflects the ineffectiveness of attrition warfare in counterinsurgency. Thompson contends that targeting the infrastructure of an insurgent group will indirectly defeat their physical order of battle, but targeting their order of battle will not defeat the infrastructure. By extension, if the infrastructure is never defeated then the order of battle may regenerate.[48] This illustrates the fact that firepower can only guarantee a gain in terms of short-term security, absent of a complimentary attempt to isolate the insurgent force from its infrastructure or hold the newly-cleared area.

Kitson

Army officer and theorist General Sir Frank Kitson[49] generally views insurgency in the same manner as Thompson, and makes a similar delineation between subversion and armed insurgency.[50] He states that an insurgent's goal is to overthrow the government, or at least force it to do something it otherwise would not do. In recognition of this and the need to oppose an insurgency with legitimate force, Kitson contends that the counterinsurgency effort is never wholly military nor wholly political.[51] In his later writings, Kitson invokes Mao's views of insurgency as an initial manifestation of total war. He describes insurgency and counterinsurgency as steps on a ladder, with a bottom step of "subversion" and a top step of "all-out war."[52]

Kitson sees the larger themes of insurgency much in a similar vein as Thompson, but tends toward operational aspects of counterinsurgency rather than the strategic aspect. Whereas Thompson focuses on principles for counterinsurgency, Kitson frames his theories in a framework of four requirements for successful campaigns:

1. Coordination at every level for the direction of the campaign.

2. A frame of mind in the population which rejects insurgent activity.

3. An effective intelligence organization.

4. Government adherence to a legal system that is suitable to the needs of the moment.[53]

Kitson's focus on operational and tactical techniques has the solid foundation of his experience in Kenya, Malaya, Aden, Oman, Cyprus, and Northern Ireland. He provides a detailed framework for the organization and leverage of tactical intelligence. Kitson asserts that there is a difference between intelligence on the general situation, and intelligence which is used to gain contact with the enemy. In this model, he contends that tactical commanders have the responsibility to develop this usable information.[54] Kitson also developed the pseudo-gang technique for infiltrating insurgent groups while posing as their confederates.[55]

Kitson's views of counterinsurgency operations have an impact on the employment of artillery units. He broadly categorizes operations somewhat classically, into offensive and defensive forms. In offensive operations, security forces attempt to identify and neutralize insurgents in the area.[56] This is in contrast to Thompson's preference of attacking the insurgent's infrastructure. Kitson also reflects that units are generally successful by assuming risk in other areas while concentrating forces in a single area until it is secure and pacified.[57] An artillery unit's indirect fires are useful in an enemy-focused operation such as Kitson describes, especially if there is a concentrated and determined enemy. His defensive operations consist of guarding key infrastructure from the insurgents and protecting the population from their influence.[58] If a counterinsurgency force is primarily engaged in these defensive operations, they may be more likely to employ artillery units in a nonstandard security role due to the implied manpower requirements.

The French Experience and Theoretical Roots

As with the British experience, the French colonial experience and military structure shaped counterinsurgency doctrine. France's colonial military operations gave rise to the theories of Lyatuey and Gallieni, who were somewhat isolated in Africa and Asia and did not develop military theories in parallel with the large standing armies in Europe. These colonial policies contended that economic attraction would be a means to co-opt a rebel's base of support. This was combined with the practice of the *razzia,* or large punitive campaigns.[59] The French had a division between the colonial military and the metropolitan infantry in Europe, which meant that many of the same officers fought in counterinsurgency campaigns completely separate from the larger institutional army.

Born of the efforts and defeats in Indochina and Algeria, modern French counterinsurgency theory also has an ideological influence acting on it. *Guerre revolutionnaire* was the unofficial doctrine of many French officers that they adopted after the perceived political sellout in Indochina, and influenced the theory that emerged from the next conflict in Algeria. This theory asserted that the strength of communist insurgencies was their ability to hyper-organize the population. In order to combat this, the French nation as a whole would have to holistically counter-organize. Seen as the guarantor of French society, the army could take responsibility of all national means if needed. *Guerre revolutionnaire* described that communist insurgencies were part of a larger global effort against western democracies. Furthermore, *guerre revolutionnaire* implied that insurgencies could not be won via negotiation.[60]

Trinquier

Army officer and theorist Colonel Roger Trinquier's[61] theories provide an excellent view of counterinsurgency through the lens of *guerre revolutionnaire*. As a veteran of colonial infantry forces in Indochina and Algeria, Trinquier's works show his emphasis on mobilizing the whole of society to fight an insurgency. He approaches insurgency as an entire clandestine organization instead of a group of individual armed elements which the government can address locally.[62] Trinquier views terrorism as a tool within insurgency, which aligns with this chapter's definition of insurgency as a form of warfare through combined military and political means. He states that the goal of an insurgency is control of the populace, so terrorism is an effective weapon because it is aimed directly at the population.[63] He views the population as the center of the conflict, and his methods seek to tightly organize it through capable local leaders to act as a fence against insurgents' penetration and the effects of their terrorism.[64] Trinquier envisions using this organized population as a means of identifying insurgents, by training select locals to act as an "intelligence action" organization. This reinforces the theories of Thompson and Kitson by implying that it is more efficient to conduct counterinsurgency through the minority of the population which supports the government, instead of trying to attract the uncommitted middle portion.[65]

Trinquier's theories of interrogation are objectionable to most military leaders when viewed through the current norms of military conduct. His discussion of physically harsh interrogation techniques comprise an extremely small portion of his writing, and they do not obviate his larger views on the nature and operational model of counterinsurgency warfare.[66] The extreme nature of his methods alone have caused some critics to impetuously question if the entire French experience in Algeria should be omitted from the current discussion on counterinsurgency.[67]

Trinquier's operational model for counterinsurgency affects both the role and distribution of artillery units. In spatial terms, Trinquier sees the counterinsurgency battlefield as three sets of areas: towns and urban centers, the inhabited rural area, and the insurgents' refuge area.[68] Trinquier designates grid troops to conduct the policing, control and organization of the population. The grid troops provide a presence within the towns and urban centers by living and working among the population. This constitutes a means of area defense. These troops have the lowest need for training in small unit maneuver tactics, since they are primarily intended to enforce resource and movement control measures.[69] As this leaves the areas between the towns and urban centers vulnerable, Trinquier designates interval units that attack the insurgents' political and military structure in that open space. He cites the need for the interval units to be composed of excellent, well-trained troops since they will be conducting primarily offensive operations.[70] To attack the insurgents in the refuge areas, Trinquier describes intervention units to conduct search-and-destroy operations in isolated regions. Since these units are likely to conduct raids and other forms of offense against the insurgency's strongest military points, Trinquier calls for only the most elite troops to constitute the intervention units.[71]

When commanders assign artillery units to a nonstandard role, they are more likely to serve as the grid troops in Trinquier's operational model. They will have a relatively lower level of training in small unit maneuver tactics compared to infantry, armor, or cavalry units. When commanders retain their artillery unit as an indirect fire provider, they must be conscious of the need to support geographically static forces (the grid troops and the interval units) as well as mobile reactive forces (the intervention units).[72]

Galula

Army officer and theorist Lieutenant Colonel David Galula[73] provides a complimentary approach to counterinsurgency from the French perspective. Trinquier focuses on protecting the population and using them to penetrate insurgent forces, while Galula focuses on using the population to build a counter-organized effort against them. Galula summarizes his counterinsurgency approach as the effort to "build (or rebuild) a political machine from the people upward."[74] To this end, he offers four laws of counterinsurgency:

1. The support of the population is equally important to both sides.

2. Support is gained through an active minority of the population.

3. Population support is conditional on who is perceived as the eventual victor.

4. The intensity of efforts and vastness of means are essential.[75]

Within this context, Galula defines victory in terms of isolating the insurgents provided that the mechanisms to isolate them are enforced by the population altruistically.[76]

Operationally, Galula divides insurgency into a cold revolutionary war when subversive actions are open and within the legal framework, and a hot revolutionary war when actions are violent and illegal. Galula's spatial model describes red areas under direct insurgent control, pink areas which are openly contested, and white areas which are under nominal government control but already have undetected subversive elements.[77] His operational model is an eight-step process which can be loosely defined as a clear, hold and build methodology. The first step of destroying insurgent forces serves to clear an area, the second step of deploying a static unit serves to hold the area, and the remaining six steps build that area's political capacity.[78]

The impact on the employment of artillery units comes from this eight-step model. Galula asserts that the population must be directed and organized since their support is seldom spontaneous.[79] He describes a myriad of tasks during the six steps of the building phase which are not necessarily the responsibility of the static unit holding the area. In this large requirement for manpower, commanders may look to their artillery unit for the manpower and command node to coordinate these tasks. During this phase, Galula cites the requirement for "some field artillery for occasional support" in recognition that the holding effort will require indirect firepower.[80] However, the commander's artillery unit and his enablers of mobile high-intensity combat may be assigned a role when their printers are more critical than their howitzers, pediatricians more valuable than forward observers, and concrete more useful than high-explosive munitions.[81]

The American Experience and Theoretical Roots

For all its modern experience in counterinsurgency, the American military has a surprising paucity of theory and doctrine on the subject. The American political and military traditions include decades of pacification in the American west, a successful counterinsurgency effort in the Philippines, and the institutionally traumatic experience of Vietnam.[82] Still, no American theorists emerge from this with the cachet of a Thompson, a Kitson, a Trinquier, or a Galula. However, one lesser-known theorist provides a detailed insight into the nature of organizing a counterinsurgency.

McCuen

Army officer and theorist Colonel John McCuen[83] provides a meticulous and prescriptive theory for counterinsurgency. Based on his experiences witnessing insurgencies in Malaya and the Philippines and advising forces in Vietnam, he shows how a hyper-organized government effort can counter a hyper-organized Maoist insurgency. His model is still valuable today because it illustrates the benefits of understanding an enemy's approach and counteracting it with all available bases of power. McCuen's theory is tailored to a host-nation government's counterinsurgency effort, although he contends that his strategic objectives and tactical actions could transfer to another element of counterinsurgency forces.[84]

McCuen defines an insurgency in four progressive phases: organization, terrorism, guerrilla warfare, and mobile warfare. As with Giap, he recognizes that insurgents in a different area may be in different phases simultaneously.[85] McCuen asserts that the optimal solution is to counter-organize along the same lines as the insurgency, which he refers to as "the application of its strategy and principles in reverse."[86] To accomplish this, a counterinsurgency force must establish strategic bases and begin to reverse the insurgency's momentum and send them backwards through the four phases. McCuen illustrates that this effort starts with determining which phase the insurgency is in, then developing a strategy to secure bases while frustrating the insurgents' attempts to establish their own bases. His method of securing a strategic base is to mobilize the population, then secure it to prevent insurgent retribution or penetration.[87] Once strategic bases are secure, counterinsurgents must seize the initiative and organize a protracted effort to roll the insurgency back to a manageable state of organization.[88]

The employment of artillery units is affected by a second-order effect of McCuen's theories. Since his operational methods rely on mobilizing the population, McCuen stresses that people must be persuaded to see that "the troops are friends and protectors rather than outsiders and destroyers."[89] This notion may lend itself to restrictions on the use of indirect firepower.

Recurring Counterinsurgency Themes

These five counterinsurgent theorists represent varied concepts for dealing with the problem of an insurgency. It is interesting that four notions are common to all of their theories. The first common element is the concept of a unity of effort, or at least a mechanism to harmonize effort across all governmental means. The idea implies political primacy as the coordinating force for all viable long-term efforts. This leads into the second common element, operating within a framework of legitimacy, which includes the use of minimum force required when confronting insurgents.[90] This lends itself to the notion that almost all insurgencies will be settled along political lines. The third common element is the requirement to isolate insurgents from their base of support, lest they grow or regenerate lost capacity. Finally, all five theorists describe the need to clear the armed insurgent element with force. There are no declarations that counterinsurgents can wait out their foes, no contentions that diplomacy and the redress of political grievances will satisfy every asset of the insurgency. They all describe the destruction of an armed insurgent in terms of the overt use of military force.[91]

These parallels in counterinsurgency theory are striking, especially in light of the different insurgencies, environments and strategic contexts that shaped their formulation. When reflecting on a counterinsurgency symposium in 1962, Kitson writes:

> Although we came from such widely divergent backgrounds, it was if we had all been brought up together from youth. We all spoke the same language. Probably all of us had worked out theories of counterinsurgency procedures at one time or another, which we thought were unique and original. But when we came to air them, all our ideas were essentially the same.[92]

Contemporary Themes in Counterinsurgency

The current deliberations on counterinsurgency theory consist of its emerging global nature and operational implementation. Counterinsurgency practitioner David Kilcullen identifies the Global War on Terror as a war against a global Islamic insurgency, and concludes that counterinsurgency theory provides the best approach to the conflict. Since these theories are designed to defeat an insurgency in a single country, they must be refined for global application.[93] With respect to operations, the debate on counterinsurgency focuses on the merits of population-centric and enemy-centric approaches in doctrine. Colonel Gian Gentile asserts that the population-centric approach perverts a military's ability to adapt to local requirements. Gentile refers to the population as "the prize" in this approach to counterinsurgency which he attributes to mainly to Galula. However, in most counterinsurgency theories (to include Galula's) the population is not the prize, but a means to the prize of isolating and defeating the insurgency.[94] Gentile also conflates the population-centric approach with nation-building activities, perhaps confounding some of the component activities which both activities share.[95]

In accordance with the framework of this study, there is no centricity in a proper approach to counterinsurgency. By categorizing counterinsurgency as two separate and distinct approaches, it implies a dualistic choice between black and white when a sound approach incorporates many shades of gray.[96] As a senior US Army officer remarked, "I don't think that you can differentiate between enemy-centric and population-centric. What you're trying to do is separate the population from the enemy. So I think it's two sides to the same coin . . . it's sort of a false dichotomy."[97]

Artillery Themes

Artillery units represent a relatively small element within the overall counterinsurgency, but an important one. Military forces are only one part of a full counterinsurgency effort, along with political, social, and economic components. But no element has the same instantaneous and lethal affect as artillery's indirect fires, and few other military arms represent a similar potential for employment in nonstandard security roles due to an artillery unit's array of equipment and combat-focused soldiers.

The Artillery Unit Defined

A comparative analysis requires a definition which frames the nature of artillery units and their functions. For this study, an artillery unit is defined as "a brigade-, battalion-, or battery-level force which is capable of providing lethal indirect fires by means of: integrated fire support tasked to maneuver units, a fire direction element, and assigned guns, howitzers, or launchers to deliver lethal fires."[98] The factor that distinguishes an artillery unit from another combat unit is its inherent ability to deliver indirect fires for a dedicated maneuver force, even if they are not currently performing tasks in that role. By defining an artillery unit as one that is capable of providing lethal indirect fires, the same terminology applies to units which are delivering fires or re-tasked for security missions. This is because the definition does not categorize based on the unit's current role.[99]

There are several subtle implications with this definition. First, this definition does not make a distinction whether an artillery unit is assigned or attached to a maneuver force conducting counterinsurgency operations, since the maneuver commander has the authority to employ the force as needed in either command relationship. This principle extends to the fire support personnel within an organization. Whether administratively assigned to the artillery unit or the supported maneuver unit, fire supporters will generally perform their functions in the same manner during sustained operations.[100] Within this study, the term "fire support" refers to both the function of the forward observer team which performs the technical tasks of employing fires with a maneuver force, and the fire support coordination element which performs the tactical or operational integration of fires with a maneuver force.

Additionally, this definition deliberately distinguishes between artillery units and mortar units. The mortar is a maneuver commander's indirect fire weapon system unless specifically re-tasked by his higher headquarters. Since the mortarmen are assigned to a maneuver unit in the same manner as the infantry or tankers are assigned, it is a primarily internal decision for that unit's maneuver commander to employ his mortars.

For comparative purposes, a battery generally consists of four to eight howitzers or rocket platforms which are further broken down into platoons or sections. The brigade is the largest, and the battery is the smallest echelon at which an artillery officer will serve as a commander to lead tactical action and direct the employment of the firing unit. At all echelons, the artillery unit depends on the supported maneuver unit's structure for most external sustainment and information requirements.[101]

Artillery Theory

There is an abundance of theory available for an examination of insurgency and counterinsurgency but very little theory exists regarding the use of artillery firepower, let alone the employment of artillery units in counterinsurgency. *Firepower in Limited War* by Major General Robert H. Scales provides an in-depth historical analysis, but offers little in the way of a resulting theory at its conclusion.[102] Additionally, the subject of firepower considers a much wider array of assets besides artillery, and the subject of limited war considers a much wider array of conflict than just counterinsurgency. Attempts to reflect on contemporary artillery operations are similarly lacking, with a 38-page occasional paper from the U.S. Army Combat Studies Institute representing the most thorough effort. Unfortunately, it is similar to Scales' work in that it is only an overview and covers the broad subject of Military Operations Other than War rather than counterinsurgency.[103] As evidence that contemporary artillery applications are more praxis than theory, much of the current military scholarship and writings in professional journals focus on narrow subjects within the artillery, such as integrating indirect fires in Iraq or adapting artillery to the environmental requirements of Afghanistan.[104]

In general, the evolution of technology and materials science impacts the employment of artillery more so than evolutions in doctrine or theory. As the ability to achieve effects via indirect fire increases, militaries seek new methods to integrate and exploit this capability. These capabilities are generally expressed in terms conducive to conventional maneuver warfare, which retards the process of implementing them into a counterinsurgency doctrine. This is because the use of indirect firepower generally enables maneuver in high-intensity conventional combat, whereas maneuver generally enables the use of indirect firepower in counterinsurgencies.[105]

The modern artillery unit is a result of the evolutionary process of institutional combat experience. As maneuver forces gained abilities to conduct coordinated attacks across a wider front in World War I, the German Colonel Georg Bruchmuller developed a centralized and task-oriented system to synchronize artillery fires. To a certain extent, this legacy is seen in most modern artillery units.[106] The U.S. Marine Corps experience with combined arms in World War II led to the development of a fire support coordination element, which was implemented and refined by many nations during the Korean War. This element shifted the burden of assigning firepower assets to target arrays from the artillery headquarters into a staff which directly served the maneuver force commander.[107]

The Vietnam War, which is examined in detail in Chapter Four, has a massive influence on the evolution of the modern artillery in a counterinsurgency environment. The two key virtues for artillery in Vietnam were responsiveness and coverage for a vast area of search-and-destroy operations.[108] To satisfy these requirements, the artillery de-centralized and developed an omnidirectional capability. For the first time since World War I, the artillery began to decentralize in order to support the counterinsurgency force's overall change in structure.[109] In 1966, infantry units began to operate in decentralized battalions instead of brigades in an attempt to hold areas after large search-and-destroy missions. To maintain a responsive firing capability, many artillery battalions broke down to isolated batteries in order to directly support infantry battalions.[110] In an effort to provide artillery support in every direction, units established battery-level firebases. As a result, fire direction was now focused at the battery level. Contemporarily, fire direction remained at the battalion level for conventional units assigned to Europe and consequently for the institutional army.[111]

Contemporary artillery and fire support is rich in doctrine and tactics, but does not have a guiding theory that attempts to explain the underlying reasons and benefits for its utilization. Perhaps the best attempt to satisfy this need is the US Army's functional concept for fires. This document is a broad concept of operationally adaptive fires, describing fires in offensive and defensive terms. In this concept, offensive fires pre-empt enemy action in an attempt to seize, retain, or exploit the initiative.[112] This type of offensive fires conceptually supports forces similar to Trinquier's intervention unit in counterinsurgency. Defensive fires protect friendly forces, population centers and critical infrastructure to provide persistent protection.[113] These fires seemingly support forces similar to Trinquier's grid units and interval units or the white areas in Galula's model of counterinsurgency.

Contemporary Themes in Artillery

The advance of technology is having an impact on artillery fires, and consequently the capacity and character of artillery units. Precision Guided Munitions (PGMs) represent an excellent advance in munitions accuracy, but are susceptible to the axiom that "precision munitions miss precisely."[114] This reflects the fact that a PGM will only be as accurate as the given target location. However, the proper and judicious use of PGMs allows a single artillery unit to aggregate effects across a wide range simultaneously instead of achieving a massing effect at a single point. Combined with the fact that PGMs are still relatively cost-prohibitive for universal use, they are generally reserved for prioritized targets which meet certain selection standards to isolate any target location error.[115] Artillery fires, and by extension artillery units, must be employed with the understanding that "no amount of technology or firepower will make secure a region unless the support of the population has been gained."[116] This reflects the concept that increases in technology and firepower only lend an absolute advantage in linear systems, but in a complex system like the counterinsurgency environment it is only a relative advantage. As the ensuing case studies illustrate, this relative advantage may be negligible.[117]

The Four Factors

Factors which influence the employment of artillery units in counterinsurgency provide a useful mechanism for analysis in the case studies which follow. These factors are presented as recurring themes, and are not intended to provide a predictive analysis of artillery employment in theoretical conflicts. Each emerging conflict must be carefully studied to inform the decision on how to use an artillery unit within a unique set of circumstances. These four factors engender the main considerations for utilizing artillery units, and as such they are useful in informing that decision but not strictly predictive of a successful result.[118]

Requirements for Indirect Fires

A major factor which influences the employment of artillery units is the requirement for tactical firepower. If the military environment has an abundance of targets which security forces cannot adequately affect, commanders are more likely to employ artillery units to deliver indirect fires. This may be the result of insufficient means to engage and destroy targets with security forces, or an ill-suited security framework that attempts to continually clear insurgents without holding that ground afterwards. Two dynamics contribute to the requirement for tactical indirect fires in a counterinsurgency: dispersion of the enemy and the frequency of contact.[119]

If an insurgent force is deliberately dispersed or conducting subversion activities, they will generally not mass in sufficient numbers that compel a security force to call for indirect fires. This insurgent force will be extremely difficult to detect, but most counterinsurgent forces will be able to overwhelm them with superior organic firepower if they are detected and remain engaged. In recognition of this, Kitson states that artillery fires are only needed when insurgents "openly take to the field."[120]

The frequency of contact also impacts the requirements for tactical indirect fires. In general, more frequent contact with the enemy creates a demand among security forces for more indirect fires. This is manifested in the degree of risk a commander assumes. If the insurgent enemy operates in a concentrated force for short time to make a limited attack, a robust indirect firepower capability is only beneficial during that limited timespan. In this case, the commander uses his artillery unit to hedge his risk against insufficient firepower but for the remainder of the campaign, the unit sits idle. If the commander assumes risk with respect to indirect fires during these few engagements, his benefit is the availability of the artillery unit for other tasks. This reactive capability to mass indirect fires at a critical point is distinct from the deliberate, preplanned fire missions that a relatively small firing unit can perform. As evidenced in several of the case studies which follow, some commanders strike this balance by detaching a small element from the artillery unit to provide fires while the majority engages in another task. This not only serves to hedge some of the risk associated with a concentrated enemy force, it also provides a resource to conduct pre-planned fire missions.

Constraints and Limitations on Indirect Fires

In the counterinsurgency environment there will be both constraints and limitations acting on the use of an artillery unit's indirect fires. This is not unique to counterinsurgency, since all forms of warfare exhibit constraints and limitations such as a justified Rules of Engagement (ROE). Constraints are the restrictions that a higher echelon places on a lower echelon, whether political or military. Limitations are physical restrictions on the application of indirect fires. Perceptive leaders will be able to distinguish between the two, and review constraints to ensure they are not unduly restricting the use of artillery fires.

The ROE is the chief constraint on a commander's ability to leverage an artillery unit's indirect firepower. These restrictions are equally conceptual and pragmatic. Conceptually, the use of minimum force serves to reinforce the legitimacy of the counterinsurgency effort by exhibiting restraint and the responsible application of violence. Pragmatically, commanders seek to apply force properly in order to maintain a positive image of their soldiers within the population.

Collateral damage considerations are often conflated with general ROE application, but here is an important distinction. The ROE addresses intentional conduct, while collateral damage policies address incidental effects due to a weapons system's lack of scalability or its inherent limits of accuracy. With indirect fires, this is a major concern. On a dismounted patrol, a soldier may suffer a momentary lapse of concentration and commit a negligent discharge with his rifle. In almost every case, the round will arc into the imperceptible distance or lodge itself harmlessly into a wall.[121] But when a round of high-explosive artillery strikes inaccurately, the effects are generally more severe. Collateral damage estimation is not supposed to delay immediate calls for indirect fire, but the ability to prosecute deliberate targets is restricted when estimated collateral damage exceeds the threshold set by a higher headquarters.[122] Minimizing collateral damage is important to local commanders. McCuen illustrates this fact by stating that counterinsurgents "should be very careful to avoid the shotgun approach: this is, accepting a few neutral civilian casualties to get a few rebels. It does not work this way because the few neutral casualties create ten-fold new rebels among the casualties' friends and relatives."[123]

Limitations on the use of indirect firepower largely result from the physical and demographic terrain of a given area. Galula sees terrain as so important that some insurgencies are doomed to failure before they even start.[124] Conversely, some counterinsurgent efforts suffer greatly from the fact that they do not choose the setting of the conflict either. Features such as dense jungle, urban terrain or mountainous regions cause a decrease in the effective range of indirect fire weapons systems since they must usually be fired at a high angle.[125] In an urban environment, population density may be so high that indirect fires are rarely used due to extreme collateral damage. A barren and flat terrain lends itself to the use of indirect fires, but the aggregate of any other features limits their use.

Counterinsurgent Force Organization

The overall composition of the counterinsurgency effort also has a large impact on the employment of artillery units. Metaphorically, there are forces that pull and push artillery units into nonstandard roles. A lack of sufficient security forces pulls artillery units into those roles, while the effectiveness of other firepower platforms may push them into nonstandard roles.

Having more units to conduct security operations in a counterinsurgency is almost always beneficial. With increased numbers, security forces can disperse to hold more area, or concentrate additional forces in critical areas. With increased numbers, a higher number of security forces working among the local population increases the daily contact between the two. This helps to increase situational awareness within the counterinsurgency force and prevent the reintroduction of insurgent forces. On the basis of this advantage, sometimes commanders use artillery units in a maneuver role to perform security tasks. If a counterinsurgent force is stretched thin in an economy-of-force mode, there will be a strong pull to employ artillery units as additional security. If there are already sufficient units to hold secure areas, this pull will not be as strong.

No matter which role an artillery unit performs, there are considerations of adequate force structure when converting one to an alternate role. An artillery battery is generally three-fourths the size of a sister rifle company, and has the junior leadership for two firing platoons compared to three rifle platoons. At the battalion echelon, this disparity in force structure is magnified. An artillery battalion will generally have less than half of the end strength of a sister infantry battalion, and is organized into only two firing batteries compared to the four companies in an infantry battalion.[126] Kitson recognizes this general disparity and concludes that it is necessary to combine two artillery units into one for the infantry role.[127]

Aside from the need for additional security forces, there is another strong pull for forces to conduct those tasks which the combat unit is structurally ill-suited to perform. Since the brigades and battalions in modernized states are generally structured for conventional operations, they do not have specified units that train to perform critical tasks such as unit advisers or civil coordination. When deployed as individual units to a new counterinsurgency, brigades and battalions cannot rely on units from higher echelons to perform the tasks of route security or force protection. Galula recognizes this fact and asserts that while a commander always needs "some field artillery for occasional support," a majority of the artillery force must be reorganized to face the myriad of tasks in counterinsurgency to include interdicting prohibited movement, conducting information operations, and implementing reforms.[128] Some critics go as far as contending that the field artillery branch itself should be the primary source for combat adviser teams in counterinsurgency.[129]

While the needs for additional security, advisory and other forces in counterinsurgency constitute a metaphorical pulling force on artillery units, the effectiveness and availability of complementary firepower constitutes a pushing force. If similar indirect firepower is effective and readily available from other sources, it relieves artillery units for another task. In the purest example of this, another artillery unit providing general support to the whole counterinsurgency force may fulfill the requirements for indirect fires in a relatively stable area. Likewise, in a low threat environment, a security force's own mortars may be sufficient to fulfill these requirements.

Unless another element can replicate the ability to mass indirect fires in almost any condition and remain indefinitely available to the commander, they are not a suitable alternative to an artillery unit. Although aviation assets provide excellent fire support options for maneuver forces, they are subject to the limitations of weather, payload, and loiter time. Unless an aviation asset is coordinated for a pre-planned mission, they will also be limited by their mix of munitions on board, transit time to the point of conflict, and a lack of scalable munitions. Commanders that replace the indirect fires capability of an artillery unit with that of an aviation asset assume a certain level of risk. In this instance, it is less of a force that pushes the artillery unit into another role, and more likely a case of the artillery unit being pulled into another role with external aviation assets mitigating the loss of indirect fires capability.

Artillery Unit Conversion Cost

The fourth factor that influences the employment of artillery units in counterinsurgency is the cost of converting these units to perform in an alternate role, and the cost to re-convert them to deliver indirect fires. As Kitson states in *Low Intensity Operations*, "neither the officers nor the soldiers of arms other than infantry are properly trained to act as infantry."[130] This reflects the fact that artillery units cannot simply receive a new tasking and hope for success; it requires sufficient time to train and develop small teams, leaders and collective actions. One major difficulty is the balance between training an artillery unit for work as a maneuver force in counterinsurgency while simultaneously providing indirect fires for other forces' train-up.[131]

When artillery units return to firing operations after a long-term tasking in another role, they must have an intensely focused training regimen to regain the technical competencies and procedures that ensure reliable firing operations. For this reason, the use of an artillery unit in another role must be short in duration due to the rapid deterioration of technical skills. This entails training and certification for both crews and fire direction centers. If the fire support element has not been adjusting and controlling fires from other sources, they will also require training and certification. The commander at the higher echelon must acknowledge that there is time required to train between completion of the nonstandard role and a return to firing capability.[132] Generally, a motivated unit without external taskings can accomplish this in 10 days at a secure location, even after a full year of employment in a maneuver force role.[133]

Notes

1. Carl von Clausewitz (1780-1831) was a Prussian military officer with service during the Napoleonic wars. His theories examine moral, social, and political aspects of warfare in terms of operations and strategy.

2. Peter Paret, "Clausewitz," in *Makers of Modern Strategy*, ed. Peter Paret (Princeton, NJ: Princeton University Press, 1986), 186-187.

3. Carl von Clausewitz, *On War*, ed. and trans. Michael Howard and Peter Paret (Princeton, NJ: Princeton University Press, 1976), 86-87.

4. Clausewitz, *On War*, 87.

5. Clausewitz, *On War*, 75.

6. Clausewitz's "paradoxical trinity" is often presented as having several layers. He describes the abstract trinity in *On War* as a "blind natural force; of the play of chance and probability within which the creative spirit is free to roam; and of its element of subordination, as an instrument of policy, which makes it subject to reason alone." In the following paragraph, Clausewitz continues to show how each of these forces of the trinity express themselves: the blind natural force expressed as the attitude of the people, the play of chance and probability expressed as the professional qualities of the army (due to his view of war as being inherently uncertain and unpredictable), and the subordination to policy expressed as the political aims of the government in war. The primordial violence inherent to the irrational force of the population serves to expand war past the limited political objectives. *Clausewitz*, 89; Michael Howard, Clausewitz (Oxford, UK: Oxford University Press, 1983), 20.

7. Werner Hahlweg, "Clausewitz and Guerrilla Warfare" in *Clausewitz and Modern Strategy*, ed. Michael Handel (London: Frank Cass, 1986), 128-131; Clausewitz, 479-483.

8. Baron Antoine-Henri de Jomini (1779-1869) was a Swiss military officer with service to the French and Russian forces during the Napoleonic era. A contemporary of Clausewitz, his theories focus more on the tactical prescriptions for success on the battlefield.

9. John Shy, "Jomini," in *Makers of Modern Strategy,* ed. Peter Paret (Princeton, NJ: Princeton University Press, 1986), 143, 148.

10. Jomini, Baron de., *Summary of The Art of War, or A New Analytical Compend* trans. Major O. F. Winship and Lieutenant E. E. McLean (New York: Putnam, 1854), 40, 46.

11. Jomini, *Summary of The Art of War*, 41.

12. Jomini, *Summary of The Art of War*, 42.

13. Shy, "Jomini," 164.

14 Sun Tzu was a Chinese military strategist and theorist (or alternately, a pseudonym to lend cachet to a cumulation of separate theorists) who lived sometime between the 7th and 4th centuries B.C. His writings have been widely adopted and applied to ventures outside of the military sphere, such as negotiations and corporate conduct.

15. Sun Tzu, *Manual for War*, trans. T.W. Kuo (Chicago: ATLI Press, 1989), 12.

16. Sun Tzu, *Manual for War*, 35.

17. John Shy and Thomas Collier, "Revolutionary War," in *Makers of Modern Strategy,* ed. Peter Paret (Princeton, NJ: Princeton University Press, 1986), 823.

18. Department of the Army, Field Manual 3-24, *Counterinsurgency* (Washington, DC: Department of the Army, 2006), 1-1.

19. Counterinsurgency theorists all provide alternate definitions for insurgency, which are discussed in the later section in this chapter devoted to counterinsurgency theory. David Gompert of the RAND Corporation provides another clear definition, describing an effort that will "seek to replace an existing order with one that conforms to their political, economic, ideological, or religious vision." David C. Gompert and John Gordon IV, *War By Other Means: Building Complete and Balanced Capabilities for Counterinsurgency* (Santa Monica, CA: RAND, 2008), 23.

20. Mao Tse-Tung (1893-1976) was a Chinese Marxist revolutionary leader and eventual ruler of the People's Republic of China. Mao was born to a middle-class peasant family, and found the communist cause through his studies at Peking University. Mao's writings reflect a rare combination of revolutionary leader, conventional military leader, and political leader.

21. Mao Tse-Tung, *On Guerrilla Warfare*, trans. by Samuel B. Griffith (Chicago: University of Illinois Press, 1961), 44-45.

22. Mao, *On Guerrilla Warfare*, Tables 1-4. At times, Mao's works were overly prescriptive because they also functioned as an inspirational field guide to revolutionaries who did not have the benefit of formal Marxist training. This accounts for the dual nature of his writings, which often switches from prescriptive details to broadly descriptive themes.

23. Mao Tse-Tung, *The Selected Military Writings of Mao Tse-Tung* (Peking: Foreign Language Press, 1972), 211.

24. Mao, *The Selected Military Writings of Mao Tse-Tung*, 212-214.

25. Mao, *The Selected Military Writings of Mao Tse-Tung*, 214.

26. Mao, *On Guerrilla Warfare*, 54-55.

27. General Vo Nguyen Giap (b. 1911) is a former Vietnamese military leader who distilled his insurgent theories into practice during his wars in Indochina against the French colonials and the American-backed South Vietnam. Initially attracted to Vietnamese nationalist movements, Giap joined a communist movement during protests against French rule in the 1930s.

28. Mao, *On Guerrilla Warfare*, 92-93.

29. Vo Nguyen Giap, *The Military Art of People's War*, ed. Russell Stetler (New York: Monthly Review Press, 1970), 168, 175.

30. Vo Nguyen Giap, *Inside the Vietminh: Vo Nguyen Giap on Guerrilla War* (Quantico, VA: Marine Corps Association, 1962), I-10.

31. Giap uses the terms *stage of contention, stage of equilibrium*, and *stage of counter-offensive*. Giap, *Inside the Vietminh: Vo Nguyen Giap on Guerrilla War*, I-4.

32. Giap, *The Military Art of People's War*, 179-181.

33. Giap, Inside the Vietminh: Vo Nguyen Giap on Guerrilla War, I-9-I-10.

34. This must be contrasted with the Marxist revolution in Cuba, and its resulting theory of "focoism" (*foco* referring to a mobile point of insurrection). In contrast with the Maoist model, Fidel Castro's small rebel band did not politically organize before initiating an armed struggle, relying on the momentum of the masses as they advanced to Havana, enabled by the extreme unpopularity of the ruling Batista regime. When this model was theorized by one of his lieutenants, Ernesto "Che" Guevara, he attempted to apply it in Bolivia to disastrous consequences since that population was not naturally attracted by the means of violence. Shy and Collier, 849-850; Che Guevara, *Guerrilla Warfare*, ed. Brian Loveman and Thomas M. Davies (Lincoln, NE: University of Nebraska Press, 1985), 47-48.

35. John Mackinlay, *The Insurgent Archipelago* (New York: Columbia University Press, 2009), 221-222.

36. Mackinlay, *The Insurgent Archipelago*, 7.

37. John Mackinlay and Alison Al-Baddawy, *Rethinking Counterinsurgency* (Santa Monica, CA: RAND, 2008), 56.

38. For an in-depth description of these types of insurgency, see Gompert and Gordon, 25-32.

39. For an illustration of the preponderance of US forces in these operations, see Carter Malkasian, "Counterinsurgency in Iraq," in *Counterinsurgency in Modern Warfare*, ed. Daniel Marston and Carter Malkasian (Oxford, UK: Osprey Publishing, 2010), 293-295.

40. Robert Thompson, *Defeating Communist Insurgency* (London: Chatto and Windus, 1966), 55; Frank Kitson, *Bunch of Five* (London: Faber and Faber, 1977), 283.

41. This reflects the following counterinsurgency operations: Malayan Emergency 1948-1960, Mau Mau Uprising (Kenya) 1952-1960, Cyprus 1955-1960, Aden 1963-1967, Oman and Dhofar 1965-1975, Northern Ireland 1969-2007, Afghanistan since 2001, and Iraq 2003-2009. *Counterinsurgency in Modern Warfare* contains a chapter on each of these conflicts except for Kenya and Cyprus, which Kitson describes (along with Malaya and Oman) in *Bunch of Five*. Julian Paget also describes Malaya, Kenya, and Cyprus in *Counter-Insurgency Operations: Techniques of Guerrilla Warfare*.

42. Sir Robert Thompson (1916-1992) was a British administrator and adviser, with service during the Malayan Emergency and within the British embassy during the Vietnam War from 1961 to 1965. Thompson wrote several volumes of theory and constructive critique of the Vietnam conflict.

43. Thompson, *Defeating Communist Insurgency*, 51.

44. Thompson, *Defeating Communist Insurgency*, 21.

45. Thompson, *Defeating Communist Insurgency*, 28.

46. These five principles are paraphrased from the original, found in *Defeating Communist Insurgency,* 50-58.

47. Robert Thompson, *No Escape From Vietnam* (New York: David McKay Co, 1969), 167-168. Thompson describes the military role as expanding government-controlled areas (clearing) and denying freedom of movement in rural areas (holding), although in practice it usually focuses on the destruction of main force units.

48. Thompson, *No Escape From Vietnam*, 167.

49. General Sir Frank Kitson (b. 1926) is a former military officer and theorist with service in Kenya, Malaya, Cyprus, Aden, Oman and Northern Ireland. He eventually rose to become the Commander-in-Chief of the UK Land Forces.

50. Frank Kitson, *Low Intensity Operations* (London: Faber and Faber, 1973), 3.

51. Kitson, *Bunch of Five*, 282-283.

52. Frank Kitson, Warfare as a Whole (London: Faber and Faber, 1987), 2-3, 175.

53. These four requirements are paraphrased from the original, found in *Bunch of Five*, 284-291.

54. Kitson, *Bunch of Five*, 296-297.

55. For his account on the creation of pseudo gangs in Kenya, see *Bunch of Five*, 29-41.

56. Kitson, *Bunch of Five*, 296.

57. Kitson, *Low Intensity Operations*, 133.

58. Kitson, *Bunch of Five*, 294-295.

59. For a complete discussion on the works of these French colonial theorists, see Douglas Porch, "Bugeaud, Gallieni, Lyatuey: The Development of French Colonial Warfare," in *Makers of Modern Strategy*, ed. Peter Paret (Princeton, NJ: Princeton University Press, 1986), 376-407.

60. Douglas Porch, "French Imperial Warfare, 1945-1962," in Counterinsurgency in Modern Warfare, ed. Daniel Marston and Carter Malkasian (Oxford, UK: Osprey Publishing, 2010), 94.

61. Colonel Roger Trinquier (1908-1986) was a French military leader and theorist. His theories represent some of the more extreme measures based on the concept of *guerre revolutionairre*. Trinquier's experience includes Indochina and Algeria.

62. Roger Trinquier, *Modern Warfare: A French View of Counterinsurgency* trans. Daniel Lee (London: Pall Mall Press, 1964), 9.

63. Trinquier, *Modern Warfare,* 16-19.

64. Trinquier, *Modern Warfare*, 29-34. This study uses the population whereas Trinquier's translated work refers to the inhabitant and the population

seemingly interchangeably.

65. Trinquier, *Modern Warfare*, 4, 39.

66. Trinquier, *Modern Warfare*, 20-23. Unfortunately, Trinquier states that "If the prisoner gives the information requested, the examination is quickly terminated; if not, specialists must force his secret from him. Then, as a soldier, he must face the suffering, and perhaps the death, he has heretofore managed to avoid."

67. For a recent effort to this end, see: Geoff Demarest, "Let's Take the French Experience in Algeria Out of U.S. Counterinsurgency Doctrine," *Military Review* (July-August 2010): 19-23. Dr. Demarest contends that the overall operation was a strategic disaster, so that it should be removed from US Army study and doctrine. This point does not address the fact that the French in Algeria were extremely successful militarily in the nearly complete destruction of the insurgent force even after they abandoned the use of torture. Aspects of this seem very valuable for study in a military manual. In a larger sense, failed missions have as much value as successful missions to inform the study of counterinsurgency.

68. Trinquier, *Modern Warfare*, 70-71.

69. Trinquier, *Modern Warfare*, 72-74.

70. Trinquier, *Modern Warfare*, 75-76, 90.

71. Trinquier, *Modern Warfare*, 80-86, 91.

72. These implications are supported by evidence in OIF. When artillery units were assigned in a security force role, they were generally employed as "land owners" with a geographic area of operations which Trinquier would recognize as a grid unit. Additionally, separate artillery units were deployed in direct support to units which served as the elite intervention units in theater.

73. Lieutenant Colonel David Galula (1919-1967) was a French military leader and theorist. Raised in French North African territories, Galula observed the Maoist Chinese efforts as well as the Greek Civil War and the Huk Rebellion in the Philippines. By the time he served as a company commander in Algeria, he had an urge to test his basic theories which he describes in *Pacification in Algeria*.

74. David Galula, *Counterinsurgency Warfare: Theory and Practice* (New York: Praeger, 1964), 136.

75. These four laws are paraphrased from the original, found in Galula, *Counterinsurgency Warfare: Theory and Practice*, 74-80.

76. Galula, *Counterinsurgency Warfare*, 77.

77. Galula, *Counterinsurgency Warfare*, 63-64, 70-71.

78. Galula devotes Chapter Seven of *Counterinsurgency Warfare: Theory and Practice* to describing the operational and tactical methods of this model.

79. David Galula, *Pacification in Algeria*, 1956-1958 (Santa Monica, CA: RAND, 1963), 114.

80. Galula, *Counterinsurgency Warfare: Theory and Practice*, 93.

81. Paraphrased from Galula's description of a converted infantry force into

a role supporting the building phase. Galula, *Counterinsurgency Warfare: Theory and Practice*, 94.

82. The dual American experiences of policing vast western territories for Native American uprisings and quelling the insurgencies of a pseudo-empire had a brief overlap in 1898. Andrew J. Birtle, *U.S. Army Counterinsurgency and Contingency Operations Doctrine 1860-1941* (Washington, DC: Center of Military History, 1997), 99.

83. Colonel John J. McCuen (1926-2010) was an American Army officer and theorist, with experience as an adviser in Vietnam. He continued to serve as an instructor on counter-guerrilla operations at the U.S. Army War College.

84. John J. McCuen, *The Art of Counter-Revolutionary War* (Harrisburg, PA: Stockpole Books, 1966), 21.

85. McCuen, *The Art of Counter-Revolutionary War*, 40-44.

86. McCuen, *The Art of Counter-Revolutionary War*, 77.

87. McCuen, *The Art of Counter-Revolutionary War*, 54-58.

88. McCuen describes these four steps in a short chapter, "The Solution-Counter-Revolutionary Strategy." McCuen, *The Art of Counter-Revolutionary War,* 77-80.

89. McCuen, *The Art of Counter-Revolutionary War*, 62.

90. This notion is concurrent with the concept that counterinsurgencies are in large part a competition in government. As such, the "legitimacy" of the counterinsurgency effort must be in the terms of the local population, not the central government or any intervening forces. This can include some extremely draconian and coercive methods, as Chapter Three discusses in depth. Additionally, this concept is not a values-based moral imperative, it is a pragmatic attempt in most counterinsurgency literature. For a good introduction on the effects of legitimacy and the Rule of Law fostering bottom-up reform from the local level in a counterinsurgency, refer to David Kilcullen's chapter titled "Deiokes and the Taliban" in *Counterinsurgency*. David Kilcullen, Counterinsurgency (Oxford, UK: Oxford University Press, 2010), 147-161.

91. The task of clearing insurgents with armed force pervades all other tasks for the counterinsurgency effort, although this is not the military forces' sole mission.

92. Kitson, *Bunch of Five*, 200. Kitson refers to a symposium directed by President Kennedy in April, 1962 which was conducted by the RAND Corporation. Stephen Hosmer and Sibylle Crane, *Counterinsurgency: A Symposium 16-20 April 1962* (Santa Monica: RAND, 1962), www.rand.org/pubs/reports/R412-1.html (accessed 30 May 2011).

93. David Kilcullen, *Counterinsurgency* (Oxford, UK: Oxford University Press, 2010), 166.

94. Gian P. Gentile, "A Strategy of Tactics: Population Centric COIN and the Army," *Parameters* (Autumn 2009): 5.

95. Gentile, "A Strategy of Tactics: Population Centric COIN and the Army,", 15.

96. To paraphrase David St. Hubbins from the 1984 movie *This is Spiñal Tap*: the two approaches are like fire and ice, and the proper approach is to be kind of the middle of that, kind of like lukewarm water. Michael McKean, *This is Spiñal Tap*. Directed by Rob Reiner. Wilmington, NC: DiLaurentis/Embassy Pictures, 1984).

97. BA010, Brigade Commander, Interview by Richard Johnson and Thomas Walton, Fort Leavenworth, KS, 22 February 2011.

98. US Army Field Manual 1-02, *Operational Terms and Graphics,* defines fire support as "fires that support land, maritime, amphibious, and special operations forces to engage enemy forces, combat formations, and facilities in pursuit of tactical and operational objectives."

99. US Army Field Manual 1-02, *Operational Terms and Graphics*, defines fires as "the effects of lethal or nonlethal weapons," and indirect fire as "fire delivered on a target that is not itself used as a point of aim for the weapon or the director." Although artillery units may also engage targets with direct fires given certain conditions, their core competency of delivering indirect fires is the distinguishing characteristic for this study. For an overview of the overall warfighting function of "Fires," see Department of the Army, Field Manual 3-0, *Operations* (Washington, DC: Department of the Army, 2008), 4-4.

100. There is an enduring argument within the military regarding the best method of assigning fire support personnel. Proponents of assigning them with the maneuver force cite the need to develop cohesive teams, while proponents of assigning them with their parent artillery unit assert that this ensures standardization, more responsive fires and senior artillery leader mentorship.

101. As of the writing of this study, the current US Army doctrine which describes the echelons of artillery units and their command support relationships was published in 2001, before the transformation to modular BCTs. For this description, see Department of the Army, Field Manual 3-09.21, *Tactics, Techniques and Procedures for the Field Artillery Battalion* (Washington, DC: Department of the Army, 2001), 1-2. Updated doctrine is still in final draft form, with a disclaimer that "The material in this manual is under development. It is NOT approved doctrine and CANNOT be used for reference or citation. The approved FM is still current and must be used for reference, or citation, until this draft is approved and authenticated" dated 18 October 2010.

102. Robert H. Scales, *Firepower in Limited War* (Novato, CA: Presidio Press, 1995), 287-296.

103. Larry Yates, *Field Artillery in Military Operations Other Than War: An Overview of the US Experience* (Fort Leavenworth, KS: Combat Studies Institute Press, 2005).

104. Patrovick G Everett, *The Role of Field Artillery in Counterinsurgency* (Fort Leavenworth, KS: Command and General Staff College, 2005), 49-55; Joseph A. Jackson, "Howitzers on the High Ground" (Monograph, School of Advanced Military Studies, Fort Leavenworth, KS, 2009), 44-49.

105. J. B. A. Bailey, *Field Artillery and Firepower* (Oxford, UK: Oxford Military Press, 1989), 241.

106. Bruchmuller's actual methods were closely copied by the Soviets and survive in some form to this day; the general concept of a task-based artillery system is a major element of modern American fire support. For Bruchmuller's impact on American artillery, see David T. Zabecki, *Steel Wind: Colonel Georg Bruchmuller and the Birth of Modern Artillery* (Westport, CT: Praeger, 1994), 116-117.

107. Bruce I. Gudmundsson, *On Artillery* (Westport, CT: Praeger, 1993), 148.

108. Gudmundsson, *On Artillery*, 151. These qualities of responsiveness and coverage will be discussed at length in Chapter Four.

109. David Ewing Ott, *Vietnam Studies: Field Artillery, 1954-1973* (Washington, DC: Department of the Army, 1975), 42-43.

110. Boyd L. Dastrup, *King of Battle: A Branch History of the U.S. Army's Field Artillery* (Washington, DC: Center of Military History, 1992), 283.

111. Bailey, *Field Artillery and Firepower*, 240.

112. US Army Training and Doctrine Command, TRADOC Pamphlet 525-3-4, *U.S. Army Functional Concept For Fires 2016-2028* (Washington, DC: Department of the Army, 2010), 11, 23.

113. TRADOC Pamphlet 525-3-4, *U.S. Army Functional Concept For Fires 2016-2028*.

114. The origin of this saying is unknown, but it is widely used within US Army fire support circles. The author first heard it used in 2004 from a senior fire support NCO during an artillery live fire exercise.

115. John J. McGrath, *Fire for Effect: Field Artillery and Close Air Support in the US Army* (Leavenworth, KS: Combat Studies Institute Press, 2007), 162.

116. Scales, *Firepower in Limited War*, 61.

117. Linda P. Beckerman, *The Non-Linear Dynamics of War* (Science Applications International Corporation), Section 1.5.

118. As introduced in Chapter One, these four factors are the broadly descriptive aspects which influence the employment of artillery units in counterinsurgency, and are not derived from an existing theoretical or doctrinal source.

119. This description of fires as a function of the dispersion and frequency of targets is the author's own, but is illustrated by the US Army's doctrinal approach to targeting, the "decide-detect-deliver-assess" methodology. With an extremely dispersed enemy or an infrequent amount of contact, insurgent formations will rarely meet a rational threshold for target selection standards. Department of the Army, Field Manual 6-20-10, *Tactics, Techniques and Procedures for the Targeting Process* (Washington, DC: Department of the Army, 1996), 1-4, 2-7.

120. Kitson, *Low Intensity Operations*, 136.

121. Author's experience as a company-level commander during OIF 06-08 (2007) and 07-09 (2009). A majority of unintentional consequences from small arms are the result of improper target discrimination or poor marksmanship under

the stress of combat patrols, not negligent discharges.

122. Under the Law of Armed Combat, all soldiers have the inherent right to defend themselves. However, this does not alleviate the legal and moral responsibilities to minimize civilian casualties.

123. McCuen, *The Art of Counter-Revolutionary War*, 61.

124. Galula, *Counterinsurgency Warfare: Theory and Practice,* 35.

125. The issue with firing in these conditions is the *cotangent of the angle of fall*, which describes the angle at which a round will strike the target. If a target is between tall buildings, on a steep reverse slope or in extremely dense foliage, then this angle must be considerably high. For a detailed discussion on the merits of high-angle and low-angle fires in urban environments, see First Lieutenant Christopher Boris, "Low Angle Fires for MOUT," *FA Journal* (November-December 2001): 43.

126. The approved end strength for a US Army light artillery battery in a BCT is currently 94 personnel, compared to 131 in a comparable rifle company. The approved end strength in a light artillery battalion was 287 personnel, compared to 693 in an infantry battalion. These numbers exclude Forward Support Companies habitually assigned to each battalion. Refer to the US Army Force Management Support Agency, www.fmsweb.army.mil for these Tables of Equipment (access requires US DOD Common Access Card registration).

127. Kitson, *Low Intensity Operations*, 184. Kitson does not indicate who gets to command this combined unit; the author proposes a cage fight in such a circumstance.

128. Galula, *Counterinsurgency Warfare: Theory and Practice,* 93-94.

129. For the most cogently constructed argument in favor of this, see John A. Nagl and Paul L. Yingling, "The FA in the Long War: A New Mission in COIN," *FA Journal* (July-August 2006): 33-36.

130. Kitson, *Low Intensity Operations*, 184.

131. Author's experiences as a battalion operations staff officer while training a MLRS battalion for OIF I (2003) and an airborne artillery battalion for OIF 06-08 (2006).

132. Kitson, *Low Intensity Operations*, 184.

133. BI290, Battery Commander, Interview by Richard Johnson, United Kingdom, 1 April 2011.

Chapter 3
The Malayan Emergency

The answer lies not in pouring more soldiers into the jungle, but rests in the hearts and minds of the Malayan people.

— General Sir Gerald Templer

The Malayan Emergency provides an excellent example of a pragmatic and evolving approach to an insurgency, with attention to a population's legitimate grievances.[1] The British and Malayan governments' counterinsurgency efforts illustrate the utility of a coordinated framework of legitimacy, an attempt to isolate insurgents from their base of support and integrate the disaffected population into society. Throughout the Emergency, the British Army's use of artillery units demonstrated that a counterinsurgent force can achieve lethal effects through a practical, limited and de-centralized approach.

Background

One of the most influential features of the Malayan Emergency was the physical country itself. Dense rainforest jungle provided a small dismounted force almost limitless concealment and standoff from observation and direct fire weapons. Much of the year sees daily temperatures above 90 degrees Fahrenheit. A central range of mountains as high as 7,000 feet runs along the middle of the peninsula, and other regions are subject to seasonal monsoons. These factors formed a foreboding environment that the foreign intervening force took several years to appreciate.[2]

Ethnically, the country was a mix of Malays (44 percent) and Chinese (32 percent) with smaller minorities of Indians (10.5 percent), Europeans (1.5 percent), and undocumented aboriginal peoples in the deep jungle.[3] The total population of just over 5 million had urban and rural components, but one of the most influential facets of ethnic distribution was the presence of isolated Chinese squatter populations at the jungle fringes.

With respect to structured governance, the British had a limited ability to direct and harmonize operations initially. At the close of World War II, British civil administration returned many possessions that were occupied by Japanese forces. Although Chinese forces generally stood against Japanese occupation and the Malay rajahs generally cooperated with them, the British could not force them to incorporate the Chinese into society for effective governance.[4] Additionally, the British were only formally in charge in one of the nine provinces, though they maintained the authority to coordinate the foreign policy of the federation.[5] Although they did not enjoy the freedom of action associated with direct political rule in most of Malaya, the British were able to deftly employ indirect rule with a mix of military, economic, and social power.

The Insurgency

The Chinese minority's grievances fostered an environment to support a protracted insurgency, though that insurgency did not directly constitute a movement of the people. The Chinese were systematically excluded from service in security forces and administration, although they held a plurality in six provinces and formed a middle class which maintained a better standard of living than the Malayan middle class in most urban centers.[6] The Malayan Communist Party (MCP) exploited this narrative and added an element of anticolonial struggle in an attempt to gain wider appeal among Malayans of all ethnicities. The Marxist rhetoric was not a reflection of the Chinese community; in many senses it was a best-fit model to organize elements of the subversion and terrorism campaigns. Later, reports from Surrendered Enemy Persons (SEPs) indicated that recruits were not attracted on ideological grounds, and that they had to be conditioned to view their contemporary politics as an existential conflict. Conditioned recruits "not only expected that physical violence was likely to be the final arbiter, but believed that hostility and aggressiveness were characteristic of political activities."[7]

As the MCP moved from subversion to armed insurgency, they relocated from urban centers to a completely rural base in the jungle. It organized its armed component as the Malayan People's Anti-British Army (MPABA) to invoke the theme of anticolonialist struggle by harkening to the Malayan People's Anti-Japanese Army. Simultaneously, the MCP organized its support activity in the squatter settlements, the Min Yuen. Membership in both the MPABA (quickly renamed the Malayan Races Liberation Army, or MRLA) and the Min Yuen was based on MCP party membership status, not on merit or potential.[8]

The MRLA began with extremely small and limited attacks, which were initially misread by the government as simple banditry along ethnic lines. The nature of their efforts changed drastically in the public's eye on 16 June 1948 when three British planters were killed.[9] In recognition that they had rashly initiated a government response, the MCP leader Chin Peng decided to slow operations and settle in for a protracted campaign.[10] The MRLA slowly increased operations and by early 1950 they were capable of moderate but coordinated attacks, such as the attack and burning of a village which left more than 1,000 homeless refugees.[11] The MRLA targeted planters and other soft targets when they could not gain overmatch with small security-force units.[12] This is a reflection of their challenges to procure heavy weaponry, as they did not have a third party arming them. Another symptom of their limited arms was that the MRLA's jungle camps were never defensive positions and had to be heavily guarded with sentries for early warning to facilitate an escape.[13]

The Counterinsurgency Effort

Similar to the MRLA, the colonial government was not prepared for an armed conflict when it declared a State of Emergency in response to the killings of 1948. They lacked sufficient intelligence on the MCP as well as an appreciation of their inroads to the Chinese community.[14] The British narrative to their own troops borrowed from the remarks of Dato Ono bin Ja'afar, a leader of Malayan nationalism:

> Let that be fully understood by everyone . . . that what was termed in 1945 and up to the middle of 1946 as "racial trouble" between the Chinese and Malays cropped up. But let me assure my Chinese friends that that trouble was not started by the Malays; that trouble was not started by law-abiding Chinese; it was started by the very people we are fighting now.[15]

The view of counterinsurgency as a form of war may be conceptually sound, but only if it is appreciated as a form of war with different social and political limits than conventional war. In this respect, Major General C. H. Boucher's operational approach of large-scale search-and-destroy techniques was flawed from its inception. As the General Officer Commanding (GOC) Malaya, he visualized military operations as a means to sequentially eradicate small groups of Communist Terrorists (CTs) and their immediate support structure. These sweeps lacked the conceptual limits inherent to counterinsurgency operations, and security forces conducted them similar to conventional operations with shoot-to-kill practices and the use of heavy force.[16] Many junior commanders' feelings that attrition was the only way to defeat the CTs is further evidence that they did not understand the nuances of counterinsurgency warfare, nor the need to address the Min Yuen as a base of support.[17] To compound the difficulties of these clearing operations, there was no force capable of adequately securing ethnic Chinese areas since the Malayan Police force was composed of ethnic Malays and Britons recruited for their similarly ill-fated experience in Palestine.[18]

The British slowly began to address their tactical inadequacies by harnessing the institutional experience of jungle warfare in Burma during World War II. Boucher was duly impressed during a review of Lieutenant Colonel Walton Walker's "Ferret Force," which organized in July 1948 to use small-unit tactics and jungle warfare expertise. Many of the officers in the Ferret Force had combat experience with Force 136, which had served with some members of the MRLA against the Japanese occupation in World War II.[19] Although the Ferret Force would be disbanded later in 1949, Boucher and the Commander-in-Chief of Far East Land Forces General Sir Neil Ritchie decided that all incoming units from the United Kingdom should receive sufficient training in these areas. Ritchie appointed Walker with the task of establishing a Jungle Warfare School.[20]

Although units improved their tactical capabilities across Malaya, the operational approach of large-scale search-and-destroy and disjointed government efforts failed to achieve success in defeating the CT threat. Additionally, Prime Minister Clement Attlee's decision to formally recognize communist China in January 1950 had deep implications in Malaya. British planters felt that it was "like being kicked in the teeth by your best friend," and the decision created an environment where the population doubted the long-term commitment of British forces.[21]

These operational and strategic shortcomings greeted retired General Sir Harold Briggs when he was designated as the new Director of Operations, arriving in Malaya on 3 April 1950. Briggs' post was intended to unify the overall effort, but the post lacked command authority for both security forces and government operations.[22] As a retired officer, Briggs arrived in Malaya as a civilian which assuaged some of the fears that the Emergency would be overly militarized.[23] Briggs was eminently qualified for the position, since he had experience with jungle operations in as the Commander of the 5th Indian Division in Burma during World War II, and the experience of supporting an Asian government after the war.[24]

Briggs incorporated his vision with a series of measures which were intended to destroy the MRLA's base of support in the Min Yuen and the ethnic Chinese population, and incorporate a framework of coordination at all levels. In general, the "Briggs Plan" is described in the following five actions:

1. Resettling the distant Chinese into population concentrations, and regrouping Chinese laborers in protected mines and factories.

2. Strengthening local governance and administration.

3. Increasing coordination between the army, the police and local leaders.

4. Building or improving access roads to the isolated parts of the country to increase legitimate government presence.

5. Controlling cleared areas to prevent the MRLA from returning.[25]

The Briggs Plan did not directly end search-and-destroy missions initially, but it did set the stage for a competition in governance instead of an all-out civil war along ethnic lines.[26]

To increase the coordination of the counterinsurgency effort, Briggs directed the establishment of War Executive Councils at the federal, state and district level. The appropriate civilian leader headed the council at each level. Although Briggs himself initially led the Federal War Council, the High Commissioner himself took the position to lend the council more authority.[27] The councils acted as action bodies, which meant that they were manned by the unit commanders or heads of specific government services. Additionally, the councils created operations centers to coordinate continuous operations, handle routine business and disseminate information.[28] As with any endeavor to command by committee, the State War Executive Committees (SWECs) and District War Executive Committees (DWECs) engaged in much deliberation before the civilian authority could direct a mutually satisfactory decision.[29]

As the government and security forces began to implement the Briggs Plan's measures, the British army continued to evaluate the changing situation. In July 1950, the new GOC Malaya, Major General Roy Urquhart, called for a crucial conference to review tactics. The significant outcomes were the importance of debriefing patrols, the requirement for undetected insertion and movement into the jungle, and the establishment of the Malayan Scouts to engage in deep jungle operations and be resupplied by air.[30] These measures were implemented, and by the end of 1951 battalions started decentralizing into company operations. Briggs accepted that large-scale search-and-destroy operations were not working, and stressed the importance of these smaller distributed operations.[31] The importance of the Jungle Warfare School was reasserted, and beginning in 1952 company commanders attended a twice-a-year course. Incoming junior officers and NCOs attended courses as well, and units began periodic rotations to retrain on tactics. These initiatives aided the progress towards common doctrine and structured assessments.[32]

The new Conservative administration in London prompted a comprehensive assessment of the Emergency in late 1951. Oliver Lyttleton, the Secretary of State for the Colonies, reviewed operations in Malaya and concluded that there was a need for a unified commander to direct all military and civilian forces instead of a military force under one command which aided the civilian efforts.[33] Lieutenant General Sir Harold Templer was the government's choice, and he began in February 1952. In Templer's new role that combined the Director of Operations and the High Commissioner, he had the power to direct all operations whereas Briggs had to lead by consensus and was hampered by the individual components' ability to appeal his decisions. As such, Templer answered to London and could afford to make hard decisions that were initially unpopular within his command in Malaya.[34]

Templer did not come in with a sweeping new plan; instead he added guidance and additional structure to Briggs' earlier measures. On the way to brief the Prime Minister before departing for Malaya, he remarked to an aide that his three priorities would be to coordinate intelligence under one person, to reorganize and retrain the hastily expanded police forces, and to improve communications between the government and its people so they would understand the purpose of coercive measures in a larger context.[35] He envisioned counterinsurgency as a broader struggle along military, political, social, economic and even spiritual lines. To strengthen the measure of population resettlement, Templer directed the expansion of government services to make these protected villages an attractive opportunity for the Chinese squatters. Templer also sought to institute "white areas" where citizens would be free of population control measures if the area was successfully cleared and held from CTs.[36] Luckily, rising tin and rubber prices due to the Korean War provided ample funding for Templer's expanded programs, since the Emergency was completely funded via Malaya.[37]

Narratives of the Emergency commonly equate Briggs with the failed search-and-destroy approach and equate Templer with the successful hearts-and-minds approach. This assertion ignores the fact that Briggs himself directly ended such large-scale search-and-destroy operations in 1951, and that he was limited in his ability to direct actions since all entities could appeal his decisions.[38] Briggs abandoned his plan of sequentially clearing states from south to north after the first iteration in Johore proved too difficult to coordinate with his limited powers.[39] Additionally, this notion implies a dichotomy in approaches and conflates the hearts-and-minds approach with a nonmilitary focus. Templer's leadership ability and expanded powers served to galvanize the counterinsurgency effort, but he was quoted as giving Briggs due credit for the plan itself: "All I did was make it [the Briggs Plan] work, which it did in due course to the tune of some 560,000 people in about 500 new villages. And it happened under my administration because I had the powers and the unfortunate General Briggs did not."[40] Within the context of Malaya, the hearts-and-minds approach was not simply positive political policy or civic action, but the synergy of both attractive and coercive actions with military security and information operations.[41] The search-and-destroy approach and the hearts-and-minds approach were not a mutually exclusive dichotomy but were two ends to a continuous range of conflict in Malaya.[42]

One of Templer's priorities was to reorganize and retrain security forces in Malaya, especially the Special Branch.[43] Due to rapid expansion at the outset of the Emergency, the police force was undertrained, ineffective, brutal, and almost completely ethnic Malays.[44] The military also continued to improve during Templer's tenure. In late 1952, Templer countered the Urquhart Committee's contention that existing Mission Training Pamphlets were sufficient to inform unit tactics and operations, and brought back Walker to develop a manual for all forces to deal with jungle warfare. The result was *The Conduct of Anti-Terrorist Operations in Malaya*, colloquially known as the *ATOM*. Walker completed the task in only two weeks, as he was able to draw on his previous material he wrote for the Jungle Warfare School, older Indian Army manuals and his own experience in command of 1/6th Gurkha Rifles.[45]

Military units continued to decentralize into more distributed operations, concentrating on the new villages to support food-control measures and locate CT camps. This included the integration of SEPs into the counterinsurgency effort, sometimes in such a rapid and pragmatic manner that they were simply handed uniforms and assigned to British platoons for the next patrol.[46] At higher echelons, division staffs engaged in the resourcing and administrative support to smaller units, but did not direct operations since this was purview of the associated SWEC or DWEC. This led one major general to say that "as far as I can see, the only thing a divisional commander has to do in this sort of war is to go around seeing that the troops have got their beer."[47]

Military units also took the fight to the CTs deeper into the jungle. As a result of the Urquhart conferences in July 1950, Lieutenant Colonel Michael Calvert initiated the creation of a light unit trained in deep jungle operations which could be resupplied by air. As a result, the Malayan Scouts were founded, which was renamed the 22nd Special Air Service (SAS) Regiment later in the Emergency.[48] By 1955, eleven jungle forts had been established, providing the interior aboriginal population with access to medical, educational and economic means.[49] The attractive nature of the jungle forts proved much more effective than trying to regroup that specific population, and the CTs in the area were isolated form yet another source of information and supplies.[50]

Within the effort to improve government services, Templer prodded the British components of the civil service to begin a deliberate process of *Malayanizing* the government with the goal of national independence. Local elections were tied to positive behavior in white areas and took some appeal away from the MCP's anti-colonial narrative.[51] At higher levels, the SWECs and DWECs began *Malayanizing* their elements in order to aid the transition to an independent national government; Malayans now had responsibility for their decisions in the counterinsurgency fight.[52] These graduated elections at the village, district, and then state level also served to stimulate the formation and maturation of political parties.[53] As a result, the multiethnic Alliance Party won the first election in preparation for independence in 1957. The dual successes of independence and a mature multiethnic political party in power served to undercut much of the MCP's anti-colonial narrative.[54] One SAS officer who commanded in Malaya summarized this by saying that "independence won it. Chin Peng and all his men had the carpet pulled from under them."[55] The overall framework of counterinsurgency operations were not disrupted because this had been a slow and deliberate process.[56]

By 1960, the CTs were scattered and ineffective, while successive states enjoyed the improved socioeconomic environments of the white areas. With a concerned, capable and independent government firmly in place, the government declared an end to the Emergency. Much of the success that led to this victory can be traced to components with roots in the Briggs plan, but matured under Templer's tenure: the harmony of all counterinsurgent efforts, isolation of CTs from their base of support, the connection of the Chinese to effective governance, a phased and deliberate *Malayanization*, and security force reforms in the police and Special Branch.[57] Across all of these efforts, the military component slowly matured through an iterative process of incorporating best practices from previous experiences in the jungle and best practices from the contemporary fight.[58]

The Employment of Artillery Units

The Royal Artillery (RA) had the distinction of engaging in counterinsurgency efforts even before the Emergency was declared in 1948. One RAND Corporation monograph concerning the organization of counterinsurgency forces in Malaya cites the example of the 26th RA, which began escorting convoys and patrolling roads as early as 1947 using the vast stores of demobilized World War II equipment they found in Singapore.[59] This is corroborated by a veteran's memorial website, which states that the 26th RA began patrolling for CTs in November 1947.[60] The account of the 26th RA's service in a nonstandard security role in *The Royal Journal of Artillery* contends that they had been in action during the anti-bandit campaigns beginning in 1946.[61] Whatever date the field regiment actually began active patrolling may be debatable, but it is clear that their contributions as a security element predate the declaration of the Emergency in 1948. This employment as a security force would not last long, as another RAND study from 1964 suggests that by the final phase of the Emergency, the Malaya Command was "using every gunner they had at his regular trade."[62]

This was not a unique role for British gunners.[63] They were also used as infantrymen in Kenya, nearly contemporary to the Malayan Emergency.[64] However, the employment of artillery units as maneuver forces was short-lived. By late 1950, there were several RA units serving in Malaya: one troop from 1 Singapore Regt RA under the 26th Gurkha Infantry Brigade at Johore Bahru and 26th Field Regiment RA (plus one battery of 25th Field Regiment RA) under the 63rd Gurkha Infantry Brigade at Seremban.[65] In 1951, the batteries were reconstituted and then dispersed as independent two-gun troops which operated up to 465 miles apart.[66] This created a manpower-intensive force structure because it required up to 190 men per battery to keep these dispersed sections in action.[67] In 1952, the commander of the 25th RA laid down principles to ensure the efficient use of these troops. He contended that the RA should be used to either drive CTs into prepared ambush sites or harass suspected camp locations in the jungle. This was prompted by the inefficient use of the dispersed troops by maneuver forces, since they would employ them in turns rather than in a prioritized and task-oriented manner.[68]

This was the first attempt to govern the effective use of artillery in Malaya, and it would continue with the inclusion of a specific chapter on artillery support in the *ATOM*. The *ATOM* refined the roles of artillery fires as: Flushing (into ambush areas), Harassing, Destructive Shoots (against CT camps), Blocking Escape Routes, Deception, and Illumination.[69] In response to the limited success of air strikes in the Emergency, the *ATOM* also included the advantages of using artillery support:

1. All-weather capability.

2. Day and night capability without the use of a searchlight.

3. Better tactical surprise than a loud approaching aircraft.

4. Better accuracy, so it could be employed closer to friendly troops.

5. More capable of sustaining an effort over several days.

6. More flexible due to the intimate relationship of gunner and infantryman.[70]

There is ample evidence that security forces took advantage of these capabilities. Guns were generally not sitting idle; their operational tempo and firing records indicate an unending cycle of movement, occupation of a firing point, and a sustained delivery of fires. The *ATOM* set forth a planning factor of 100 rounds per day from a troop of two 25-pounder guns in an operation.[71] The record of operations from the 93rd Field Battery, 25th RA shows an astounding tempo. In January 1953 alone, the battery's troops fired 17,262 rounds from 30 different positions, travelling 1,646 miles along the way.[72] Gunner James Daniel Lamb's war diary indicates that in his eight months as a conscript, his troop travelled to 16 different positions. As the troop's driver for resupply trips, he regularly resupplied the two-gun positions with 200 rounds, generally every four days.[73]

In response to this demand for artillery fires, the force expanded in 1955 from 16 guns to 44.[74] To give the additional troops a requisite command structure and support, the 2nd RA and 48th RA joined the order of battle.[75] In another sign of the increased appreciation for indirect fires in counterinsurgency operations, in 1955 the infantry also replaced the 3-inch mortar with the 4.2-inch mortar at the battalion level.[76] This increase of indirect firepower coincides with the steady increase of its utilization, and in a review of operations in 1957 the practices of Flushing, Harassing, Deception and Illumination were credited as effective means to employ both artillery and mortar fires.[77]

Two operations illustrate the use of artillery units during the Emergency. An example of the remarkable use of artillery fires is found in Operation Nassau, a DWEC-level operation in the Kuala Langat district of Selangor. Beginning in 1954, this nine-month operation aimed to support food-denial measures and defeat several MCP branches in the nearby jungle. The operation had a contingent of 25-pounders from the 25th RA, which initially fired at suspected locations to flush CTs into prepared engagement areas. After the initial phase, the troops switched to continuous harassment fires, and finished with more than 60,000 rounds fired to compliment an additional 30,000 mortar rounds. In conjunction with effective food-denial to the CTs and the use of small patrols, the constant harassment from all available fires support assets was credited with the success of the operation. The district was declared a white area on 22 September 1955.[78]

Operation APOLLO in 1954 demonstrates a more limited role of artillery fires in a successful operation. The Pahang SWEC decided to launch an operation with the intent of supporting food-denial measures, isolating CT camps and defeating them sequentially.[79] The operational command took control of one field troop from the 25th RA and one platoon of Malayan mortars.[80] Indirect fires were used to harass CT camps in the jungle and deny terrain, but not in any other role that the *ATOM* describes.[81] Although the guns stayed longer than their initial six-week attachment, they were not credited as a major factor in the success of food-denial operations as they were in Operation Nassau.[82]

Although the military as a whole sought to *Malayanize* the security operations and create capable Malayan units, the RA did not raise an indigenous force until the last months of the Emergency. The 1st Field Battery, Federation Artillery activated in May 1960 after training by a seconded British officer.[83] Although many members of this new battery were ex-members of the 1st Singapore RA, the regiment was not used during the Emergency.[84]

The employment of artillery units directly contributed to the military and security successes in the Emergency. Whether it was serving in a nonstandard capacity during the early days when every combat arms soldier was needed for patrolling, or providing lethal fires for jungle patrols, the men of the RA are "justifiably proud of its ability to turn its hand ... to unorthodox forms of fighting for which it is not intentionally designed."[85] However, these lessons did not permeate the larger British artillery institution nor did they extend to the American employment of artillery units in Vietnam. In the *Royal Journal of Artillery,* the only article written about the RA in Malaya is an account of the 26th RA and their patrols during the early emergency. The only other mentions from 1948 to 1960 are a one-page article on the activation of the 1st Field Battery, Federation Artillery and a short book review of Richard Miers' *Shoot to Kill*.[86] Although there are plenty of articles on horse racing and ski vacations, there is not a single article devoted to firing artillery or the employment of units during the Emergency. The lessons of employing distributed artillery troops in a counterinsurgency did not permeate the institutional British Army that was focused on the conventional Cold War threat at the time, but the lessons were leveraged during operations in Borneo.[87] Although the British would adapt other aspects of the Malayan Emergency such as population resource control to other counterinsurgencies, utilizing artillery would not be one of them.[88]

This may be responsible for the inaccurate narratives found in American studies of the Emergency. One analysis conflates the use of artillery with a conventional approach by critiquing a commander's assessment that "artillery support proved invaluable for night harassing attacks and to sustain pressure over a wide area" after Operation SPRINGTIDE in 1951.[89] This is an unfounded criticism since the evidence of a "conventional approach" lies in the use of a maneuver force in a large-scale sweep instead of in a decentralized effort to locate CT camps based on intelligence.[90] Another history of the Emergency erroneously asserts that "artillery units spent more time acting as infantrymen than in serving the guns."[91] This notion influences further studies on the subject, contrary to the aforementioned evidence of the employment of artillery units. There is only one monograph from the US Army's School for Advanced Military Studies that analyzes field artillery in counterinsurgency, and it cites this as a central piece of evidence that artillery units were under-utilized in the Emergency.[92] The following section provides an analysis of the employment of artillery units in the Emergency, based on the evidence that they were utilized in a pragmatic and decentralized fashion.

Analysis

During the Emergency, government forces were able to unify their effort. Although the structure of the military and government in 1948 did not lend itself to coordinated actions, Briggs' advent of the War Executive Committee system achieved such a function. Reflecting on the counterinsurgent apparatus prior to the Briggs Plan, Thompson wrote that "we were running into trouble. No High Commissioner could handle the government, the country, deal with the Malay Sultans – and still have time to coordinate the various security forces including the police."[93] To capitalize on the British experience in other counterinsurgencies and emergencies, the military served in aid to the civil power.[94] Initially eschewed by military commanders, the legitimacy of the SWECs and DWECs is noted by a battalion commander in 1955:

> Now warfare by committee is a positive anathema to the soldier who has been trained all of his life to make up his own mind on the conduct of operations and then carry them out. But the ponderous committee system was forced on us by the fact that in Malaya the Army was acting merely in support of, and not in place of, civil administration.[95]

As with any major organizational change, the system took some time to be accepted by the forces that would carry out the councils' planned operations. Initial issues included a paucity of qualified civil and police administrators, weak intelligence and low funding.[96] Over time, the councils proved their worth by remaining action bodies instead of just coordination centers.[97] As a fringe benefit, they also served as an expedient vehicle for *Malayanization*. As the councils replaced British members with Malayans, it tied them to the conduct of the campaign and restored accountability to Malayans ahead of elections and independence.[98]

The security forces operated within a framework of legitimacy even while prosecuting some very harsh and coercive population control measures. The *Handbook to Malaya and the Insurgency* provided guidance to soldiers that even though detention did not require evidence that would stand up in a court of law, they were not operating in a police state but in a state of emergency.[99] In a larger sense, British forces understood the legitimacy of their operations since a premature withdrawal would jeopardize the security of a population that the British were legally responsible for.[100] Even while instituting tough measures such as a 22-hour curfew while halving rations in the village of Tanjong Malim, security forces applied them within the framework of established emergency regulations and eventually tied them to security in a given area.[101] As the government addressed the legitimate grievances of the Chinese minority and an independent government took over in 1957, much of the MCP's narrative lost its appeal.[102] This could not have been accomplished with a force that operated outside of legitimate mandates.

The government's forces were also able to effectively isolate the CTs of the MRLA from its base of support, the MinYuen. After the ill-framed operational approach of large-scale search-and-destroy was abandoned in favor of supporting population relocation and security, isolation became the cornerstone of security operations. Historian and career civil administrator Robert Komer[103] provides six coordinated civil programs which he credits with breaking the link between the insurgents and their support base:

1. Registration, travel control, curfews, and identification checks.

2. Resettlement of the Chinese squatter population in protected new villages.

3. Pervasive food and drug controls to deny CTs access to sustenance.

4. Accelerated social and economic development in the new villages.

5. A steady movement to self-government and independence.

6. Effective information campaigns to inform the population of the measures.

The centrally planned program of resettlement embodies these civil programs. By 1954, more than 570,000 had been resettled into new villages, and an additional 650,000 laborers of all races would be regrouped into 1,500 protected mines and plantation estates.[104] This process was made considerably easier by the fact that the relocated population was almost entirely ethnic Chinese, allowing the government to focus on one initially unpopular segment of the population and their needs.[105] Food- and-drug control programs took months to have an effect on the CTs which depended on the Min Yuen, but they invariably had an effect over time since the CTs could not cultivate rice without observation by air.[106] Although this program became a well-practiced process by military and civil administrators, initial efforts were disastrous and threatened to condemn the entire counterinsurgency effort. The coercive nature of the program had legal roots in the Emergency Regulations of 1948 and 1949, but the combination of inadequate provisions and mass punishment in the early iterations of relocation showed that "there was a substantial element of repression to the strategy as well."[107]

These efforts to isolate the insurgents through the severance of their support were complimented by an attractive policy towards the targeted population. After several initial failures, the government quickly realized that the resettled locations must be an attractive option for the Chinese. It leveraged civil programs to achieve this. Emergency Regulations did not have their intended effects until the population was consolidated with effective local governments.[108] These programs proved extremely effective as only six of more than 500 new villages were abandoned, mostly due to security issues.[109] To complement the effective reach of governance, the people in the new villages were held accountable for their own security via the Home Guard, a local security force raised from reliable citizens in the new villages. This force marginally increased security in the new villages, but more importantly the Home Guard served to empower the Chinese minority and further undercut the CT appeal as protectors. Likewise, special constables were trained to perform local security in and around the regrouped labor centers.[110]

The security forces were also able to clear areas of MRLA forces and eventually maintain them to support governance and development. While the military owes this ability to the aforementioned improvements in tactical methods, it is also attributable to the abilities of junior commanders fusing intelligence with operations in a decentralized structure.[111] The military also learned to leverage the core competencies of different organizations, which is evidenced in the success of army units patrolling to clear CTs in the jungle while the Home Guard held the new villages.[112] Finally, the military developed the ability to conduct operations in the remote aboriginal areas of the deep jungle, taking away their last option for refuge.[113] Of particular relevance to this study, the military also developed a flexible and economical use of artillery fires to support the armed clearing of insurgents.

The Emergency provides a unique opportunity to analyze the employment of artillery units in counterinsurgency because they were generally used in a maneuver security role prior to 1951, and generally used in an indirect-fires role from then until the end of the Emergency. The requirement for tactical fires and the organization of the counterinsurgency force were the two factors that had the most influence. The limitations of the physical environment were a lesser influence, and the cost of converting units themselves has little impact on their employment.

The requirement for tactical fires had a great influence on the employment of artillery units. Without an enemy that massed in sufficient strength, there was a low demand from the maneuver forces for firepower.[114] The dispersion of MRLA forces is well documented by practitioners during the Emergency. Thompson also contends that the true nature of targets in the Emergency were not always an element that lethal fires could interdict, since "[i]t's all very well having bombers, masses of helicopters, tremendous firepower, but none of these will eliminate a Communist cell in a High School which is producing 50 recruits a year for the insurgent movement."[115] In most accounts of those who actively patrolled for CTs in the jungle, they rarely massed in groups of 20 men or larger.[116] Even small patrols could overmatch these dispersed CT elements, especially since the MRLA generally lacked any firepower larger than a rifle.[117] The resulting low demand allowed artillery units to operate in smaller de-centralized units, in accordance with the priority of fires.

In addition to the smaller scale and dispersed nature of contact in the jungle, it was also infrequent. An operational evaluation assessed that it took 1,800 man-hours on patrol to gain contact with a CT element.[118] The MRLA also attacked soft targets such as British planters where there was little state security presence and no tactical communications link to coordinate artillery support.[119] This contributed to the inability to conduct deliberate targeting, and often the default for fires was a harassing role. Although it was effective in many cases, it strengthened the force that drove artillery units to operate in smaller echelons since there was no need for massed preparatory fires. Examples include:

    1. SWEC-level Operation GINGER in 1957; there were 172 known CTs in the operational area but not a single known location.[120]

    2. SWEC-level Operation APOLLO in 1954; there were 68 known CTs but only vague descriptions of their general area.[121]

    3. SWEC-level Operation SWORD in 1953; there were no deliberate targets in an operational area for three battalions; units failed to gain and maintain contact for on-call fires even though 25-pounder and 5.5-inch guns were attached.[122]

    4. In a DWEC-level operation described by Richard Miers in *Shoot to Kill*, his battalion could only use artillery to "discourage the CT from breaking out" since they had no deliberate targets.[123]

The other factor that significantly influenced the role of artillery units was the overall composition and strength of the military during the Emergency. During the early years, the military forces could only field ten battalions, which put approximately 4,000 men in the field after adjusting for staff and support units. This was definitely an understrength endeavor, since estimates are that the MCP could field 4,000 CTs itself. Although the military would deploy more units to raise its strength to more than 8,000 men in the field by 1951, the MRLA had a corresponding increase and could also field roughly 8,000 CTs.[124] In this environment, it is easy to understand why the 26th RA and other gunners were pressed into service to act as infantrymen. With the decline of the operational approach of search-and-destroy attrition warfare, the requirement for provisional infantrymen decreased and the gunners could once again support operations with indirect firepower.

Other sources of indirect firepower were not sufficient to make artillery units available for service in a nonstandard role. Although the 4.2-inch mortar was an improvement, it did not provide enough firepower or range for a commander's needs during many patrols.[125] Most commanders contended that aerial firepower was ineffective in the jungle environment, but one unit proved that this was due to inaccurate target location and the abandonment of camps. In his anecdote of an operation he called the "Yellow Pin," Richard Meirs' unit used Special Branch assets to verify the occupation of a camp, and employed close air support to kill 13 CT's in one strike.[126] However, after 1951 there was a pervasive notion in Malaya that in general, artillery was a superior alternative to aerial firepower.[127] This is also reflected in the fact that the CTs felt that they could dig a hasty foxhole and survive almost all attacks from the air.[128]

The constraints and limitations on artillery fires during the Emergency were a significant factor in the employment of artillery units. There were not many organizational constraints on the use of fires, and the rules of engagements were extremely permissive. Founded in the legitimate emergency regulations, units could use a shoot-on-sight policy in the declared black areas. However, unit commanders themselves had to weigh the use of artillery fires and its effect on the attitude and mentality of the local population.[129] Collateral damage was not a significant concern since a majority of the fighting was in unpopulated areas. To mitigate against collateral damage at the edge of the jungle, the *ATOM* recommended using patrols to deny terrain within the first kilometer of the jungle fringe.[130]

The limitations caused by the terrain in Malaya had a far greater effect. Although the mountainous nature of the interior played a minor role by adding a degree of difficulty to indirect fires, the primary factor was the jungle. The Malayan jungle had two varieties, the *ulu* (primary jungle) and the *belukar* (secondary jungle, the result of prior logging). Movement in the *ulu* is comparatively easy, but visibility is only 20 to 30 yards. Due to the mass of undergrowth, movement in the *belukar* is extremely difficult and there is effectively no visibility.[131] The effect of patrolling in the jungle was that enemy contact was generally a short burst of small arms fire, followed by pursuit in a single file. In this scenario, there was neither time nor continuous observation to call for artillery fire.[132]

The jungle had more effects than just the conduct of small patrols. Since it was a foreboding and remote area, there was little survey and maps were generally based on terrain features. This was adequate for the infantry's navigational purposes, but proper artillery survey was not leveraged until the Emergency was well underway. Discussions on survey were non-existent until the *ATOM* was published.[133] Due to the limited accuracy of maps, the *ATOM* limited predicted fires to within 500 yards of troops for 25-pounder guns and 1,000 yards for 5.5-inch guns.[134] Through experience, commanders would later be comfortable with observed fires as close as 200 yards while in the jungle.[135]

Due to limited visibility, the observation of fires was a severe challenge to effectively integrate artillery with maneuver.[136] The use of aerial observers was widespread for priority operations[137] and included in the *ATOM* as a best practice.[138] The military tried a variety of innovative approaches that included the testing of remote infrared sensors, but they proved ineffective.[139] The effect of the jungle was a limitation in the reactive quality of artillery fire support, and thus decreased the overall requirements for large artillery units to standby for massed fires. This made the use of smaller detached troops a more attractive option, since they could be dispersed to a wider area with less overall resource requirements.

The cost of converting artillery units had a negligible effect on their employment, since they did not change roles back and forth between indirect fires and nonstandard missions after the overall shift in 1951. As soldiers in a combat arm, the gunners of 26th RA had ample training in tactical mounted movement and were able to embrace the transition to patrolling in armored cars prior to 1948.[140] Their main challenge was to train their cooks, mechanics and administrative personnel as infantrymen. As with other artillery units, they were comparatively smaller and needed every man available as a quick-reaction force.[141] Throughout the later years of the Emergency, gunners needed little retraining on core competencies since they were generally firing artillery.[142] The 25th RA advocated a periodic retraining and certification period of 10 to 14 days for each six weeks a troop spent supporting priority operations.[143] This was a reflection of the British Army's commitment to both initial and periodic retraining in Malaya.[144]

Conclusion

The Malayan Emergency showed the effectiveness of an approach to isolate insurgents from their base of support and integrate the disaffected population into society. The British and Malayan counterinsurgency effort was able to skillfully co-opt certain parts of the insurgent appeal and defeat the rest through an application of both attractive and coercive measures. Although the Malayan Emergency was not an archetype of flawless counterinsurgency practice, it does provide an illustration of what is possible through steady adaptation and enduring social and political solutions.

This led to the employment of artillery units in a manner which supported an eventually sound strategy, instead of providing fires as a means unto itself or additional troops for an ill-suited approach. The next chapter will examine their employment in a role that focused almost exclusively on attritting enemy formations through indirect firepower as an extension of a different strategic concept. Although the terrain, population, and insurgent organization in Vietnam would be immediately recognizable to a veteran of the Malayan Emergency, the scope and use of artillery would not.

Notes

1. In this chapter's epigraph, Templer is indirectly quoted in Robert Komer, *The Malayan Emergency in Retrospect: Organization of a Successful Counterinsurgency Effort* (Santa Monica, CA: RAND, 1972), 54. Although the term "hearts and minds" arguably appears as early as the Sandeman enterprises in modern-day Afghanistan, the term entered the public lexicon during the Malayan Emergency. For further research into the Malayan Emergency, some of the most helpful resources are as follows: for an in-depth background on the social and political components of the MCP's insurgency, see Lucian Pye's *Guerrilla Communism in Malaya;* for an overview of the counterinsurgency effort, refer to Richard Stubbs' "From Search and Destroy to Hearts and Minds: The Evolution of British Strategy in Malaya 1948-60" in *Counterinsurgency in Modern Warfare* and *The War of the Running Dogs: How Malaya Defeated the Communist Guerrillas 1948-1960* by Noel Barber; for studies into the British military's efforts to adapt during the Emergency, see Daniel Marston's "Lost and Found in the Jungle" in Hew Strachan's *Big Wars and Small Wars*; Raffi Gregorian's 1994 article "Jungle Bashing in Malaya" in *Small Wars and Insurgencies*; and John Nagl's *Learning to Eat Soup With a Knife: Counterinsurgency Lessons From Malaya and Vietnam;* outstanding personal accounts by tactical leaders in the counterinsurgency include Frank Kitson's *Bunch of Five* and Richard Miers' *Shoot to Kill*; and finally Riley Sunderland's series of RAND reports and Robert Komer's *The Malayan Emergency in Retrospect: Organization of a Successful Counterinsurgency Effort* provide the first sets of analysis for the conflict.

2. For detailed description of the terrain and its effect on military operations, see Raffi Gregorian, "Jungle Bashing in Malaya," *Small Wars and Insurgencies* 5, No. 3 (Winter 1994): 344-346. As described by an artilleryman with first-hand experience, see T. W. Daniel, "Man Hunting in Malaya," in *The Royal Journal of Artillery* 77, no. 3 (July 1950): 257.

3. Different sources cite different percentages of the Malayan population along ethnic lines. These figures are drawn from Richard Stubbs, "From Search and Destroy to Hearts and Minds: The Evolution of British Strategy in Malaya 1948-60" in *Counterinsurgency in Modern Warfare*, ed. Daniel Marston and Carter Malkasian (Oxford, UK: Osprey Publishing, 2010), 102-103, 106.

4. Malaya I Panel Discussion, US Army Command and General Staff College Art of War Scholars Seminar, 4 January 2011, Fort Leavenworth, KS.

5. Malaya I Panel Discussion.

6. James Corum, "Training Indigenous Forces in Counterinsurgency: A Tale of Two Insurgencies," Strategic Studies Institute, www.strategicstudiesinstitute.army.mil/ pubs/display.cfm (accessed 28 March 2011), 20.

7. Lucian W. Pye, *Guerrilla Communism in Malaya* (Princeton, NJ: Princeton University Press, 1956), 168. Additionally, Kitson describes these efforts as "fraudulent" to the legitimate Chinese population. Frank Kitson, *Bunch of Five* (London: Faber and Faber, 1977), 73.

8. Pye, *Guerrilla Communism in Malaya*, 87-91.

9. Stubbs, "From Search and Destroy to Hearts and Minds," 101.

10. Kitson, *Bunch of Five*, (London: Faber and Faber, 1977), 73.

11. John Coates, *Suppressing Insurgency* (Boulder, CO: Westview Press, 1992), 80.

12. Richard Miers, *Shoot to Kill* (London: Faber and Faber, 1959), 164. Throughout his first-hand account as a battalion commander, Miers describes several attacks on British rubber planters and other isolated vestiges of the pseudo-colonial structure in Malaya.

13. Raffi Gregorian, "Jungle Bashing in Malaya," *Small Wars and Insurgencies* 5, No. 3 (Winter 1994), 344.

14. Stubbs, "From Search and Destroy to Hearts and Minds," 101-102.

15. Federation of Malaya, *Handbook to Malaya and the Insurgency* (Singapore: Regional Information Office for the United Kingdom in South East Asia, 1953), 19. The British government used the term "emergency" instead of "war" only to ensure that it did not cause a contractual increase on insurance premiums for Malayan enterprises back on the London insurance markets. Refer to the author's notes in Noel Barber, *The War of the Running Dogs: How Malaya Defeated the Communist Guerrillas 1948-1960* (London: Cassell, 1971), 114.

16. Stubbs, "From Search and Destroy to Hearts and Minds," 103-106.

17. For example, a battery commander from the 26th Royal Artillery Regiment (which was acting as a mounted security force) recounted that "Come what may, this 'hard core' will never accept terms, and until it has been eradicated by process of attrition the present troubles will continue." Refer to Daniel, "Man Hunting in Malaya," 257.

18. Daniel, "Man Hunting in Malaya," 104.

19. Daniel Marston, "Lost and Found in the Jungle," in *Big Wars and Small Wars*, ed. Hew Strachan (London: Routledge, 2006), 97-98.

20. Marston, "Lost and Found in the Jungle," 98-99.

21. Barber, *The War of the Running Dogs*, 113.

22. Coates, *Suppressing Insurgency*, 80-81.

23. Barber, *The War of the Running Dogs*, 114.

24. Coates, *Suppressing Insurgency*, 81. Additionally, Briggs had experience in World War I in France, as well as Mesopotamia, Palestine, and India which endowed him with an appreciation for both conventional warfare and operations in aid to the civil power.

25. Stubbs, "From Search and Destroy to Hearts and Minds," 107. The list is paraphrased from Stubbs' description of the Briggs Plan.

26. Malaya I Panel Discussion, US Army Command and General Staff College Art of War Scholars Seminar, 4 January 2011, Fort Leavenworth, KS.

27. Robert Komer, *The Malayan Emergency in Retrospect: Organization of a Successful Counterinsurgency Effort* (Santa Monica, CA: RAND, 1972), 27.

28. Komer, *The Malayan Emergency in Retrospect*, 28-29.

29. For a Brigade commander's account of SWECs in action (which could be facetiously referred to as a "a SWEC's inaction"), see M.C.A. Henniker, *Red Shadow Over Malaya* (London: William Blackwood and Sons, 1955), 61-70.

30. Marston, "Lost and Found in the Jungle," 101.

31. Marston, , "Lost and Found in the Jungle," 101

32. Marston, "Lost and Found in the Jungle," 102.

33. Stubbs, "From Search and Destroy to Hearts and Minds," 108.

34. Stubbs, "From Search and Destroy to Hearts and Minds," 108-109.

35. Leon Comber, *Malaya's Secret Police 1945-1960: The Role of the Special Branch in the Malayan Emergency* (Melbourne, AU: Monash University Press, 2009), 178.

36. Stubbs, "From Search and Destroy to Hearts and Minds," 109-110.

37. Stubbs, "From Search and Destroy to Hearts and Minds," 109. Earlier in the Emergency, government forces were having issues in just meeting the payroll demands.

38. Malaya I Panel Discussion, US Army Command and General Staff College Art of War Scholars Seminar, 4 January 2011, Fort Leavenworth, KS.

39. Coates, *Suppressing Insurgency*, 94. One consideration for this abortive attempt in 1951 is the proximity of Johore to a large ethnic Chinese community across the border in Singapore.

40. Comber, *Malaya's Secret Police 1945-1960*, 174. Templer is quoted in a letter dated 1977. See Comber's endnote No. 9 for the complete context of the note.

41. Malaya II Panel Discussion, US Army Command and General Staff College Art of War Scholars Seminar, 7 January 2011, Fort Leavenworth, KS. For an additional discussion on the synergy of political, military and information operations, see Riley Sunderland, *Winning the Hearts and Minds of the People: Malaya 1948-1960* (Santa Monica, CA: RAND, 1964).

42. Stubbs, "From Search and Destroy to Hearts and Minds," 115-118.

43. Comber, *Malaya's Secret Police 1945-1960*, 181 and 183. In most security forces with a British heritage, the Special Branch is a police organization tasked with the collection of intelligence against banned political parties or other organizations that pose a subversive or terrorist threat to the government. While many may try to compare them to the American FBI or Secret Service, there is no direct counterpart in American law enforcement.

44. Templer brought Colonel Arthur Young to oversee this overhaul of the police forces. Young aimed to turn the police from a force to a service by restoring discipline, and enhanced the Special Branch's status by separating it from the other police functions. Under the direction of Guy Madoc, the Special

Branch would begin to mature into one of the government's most reliable intelligence sources. Templer also directed the close coordination between the special branches of Malaya and Singapore, which also supported his priority of coordinating intelligence. The success of these efforts was evident by 1955 when the Special Branch Training School was widely regarded as the best of its kind in Southeast Asia. Comber, 173-217.

45. Martson, "Lost and Found in the Jungle," 103.

46. AA1009, Retired General Officer, Interview by Ken Gleiman, Brian McCarthy, Travis Molloere, and Carrie Przelski, United Kingdom, 4 October 2010.

47. Komer, *The Malayan Emergency in Retrospect*, 48.

48. Martson, "Lost and Found in the Jungle," 102.

49. Coates, *Suppressing Insurgency*, 93.

50. BI070, Retired General Officer, Interview by Mark Battjes, Ben Boardman, Robert Green, Richard Johnson, Aaron Kaufman, Dustin Mitchell, Nathan Springer, and Thomas Walton, United Kingdom, 30 March 2011.

51. Stubbs, "From Search and Destroy to Hearts and Minds," 114.

52. Komer, *The Malayan Emergency in Retrospect*, 32.

53. Komer, *The Malayan Emergency in Retrospect*, 66.

54. Stubbs, "From Search and Destroy to Hearts and Minds," 114-115.

55. BI070, Retired General Officer.

56. BI070, Retired General Officer.

57. Malaya II Panel Discussion, US Army Command and General Staff College Art of War Scholars Seminar, 7 January 2011, Fort Leavenworth, KS.

58. Marston, "Lost and Found in the Jungle," 106.

59. Riley Sunderland, *Organizing Counterinsurgency in Malaya, 1947-1960* (Santa Monica, CA: RAND, 1964), 17.

60. The Virgin Soldiers, "British and Commonwealth Units That Served in the Malayan Emergency," www.britains-smallwars.com/malaya/reg.html (accessed 6 June 2011). The veterans' descriptions of the order of battle corroborates the order of battle cited later in E.D. Smith's work.

61. Daniel, "Man Hunting in Malaya," 258.

62. Riley Sunderland, *Army Operations in Malaya, 1947-1960* (Santa Monica, CA: RAND, 1964), 243.

63. The colloquial term for a British artillery soldier is gunner, analogous to the American use of artilleryman. Therefore, both terms will be used in this study accordingly, based on nationality.

64. J. M. Forster, *A Comparative Study of the Emergencies in Kenya and Malaya* (Operational Research Unit Far East, 1957), 64.

65. E. D. Smith, *Counter-Insurgency Operations 1: Malaya and Borneo*

(London: Ian Allan, 1985), 12.

66. Sunderland, *Army Operations in Malaya, 1947-1960,* 235-236.

67. Sunderland, *Army Operations in Malaya, 1947-1960,* 237.

68. Sunderland, *Army Operations in Malaya, 1947-1960,* 238.

69. FARELF Training Center, *The Conduct of Anti-Terrorist Operations in Malaya* (St. Petersburg, FL: Hailer), Chapter 18, Section 5. The final version of the *ATOM* was published in Malaya in 1958. It was recently re-printed in the original form by Hailer Publishing.

70. FARELF Training Center, *The Conduct of Anti-Terrorist Operations in Malaya*, Chapter 18, Section 3.

71. FARELF Training Center, *The Conduct of Anti-Terrorist Operations in Malaya*, Chapter 18, Section 7.

72. Sunderland, *Army Operations in Malaya, 1947-1960,* 241.

73. James Daniel Lamb, "25th Field Regiment in Malaya," www.britains-smallwars.com/malaya/malaya-25field%20reg/25_fieldregt.html (accessed 6 June 2011). The war diary of Gunner James Daniel Lamb, F Troop, 93d Field Battery, 25 RA. This account was posted online at a veteran's memorial website, and it also contains over 100 of his photos showing the daily operations of a decentralized troop in the Emergency.

74. Sunderland, *Army Operations in Malaya, 1947-1960,* 238.

75. Smith, *Counter-Insurgency Operations 1*, 107.

76. Sunderland, *Army Operations in Malaya, 1947-1960,* 238.

77. Forster, *A Comparative Study of the Emergencies in Kenya and Malaya*, 65.

78. Sunderland, *Army Operations in Malaya, 1947-1960,* 152-155.

79. The State Secretariat of Pahang, *Pahang SWEC Operation No. 2: Operation Apollo* (1954), 1-2.

80. District Office of Kuala Lipis, *Lipis DWEC Operation No. 2: Operation Apollo* (1954), 2.

81. District Office of Kuala Lipis, Final Operation Report.

82. District Office of Kuala Lipis, Annex A.

83. G. C. K. Rowe, "Presentation of a Silver Centre-piece to the Federation Artillery," *The Royal Journal of Artillery* 87, no. 3 (Summer 1960): 128-129.

84. Rowe, "Presentation of a Silver Centre-piece to the Federation Artillery," 129; Smith, *Counter-Insurgency Operations 1,* 107.

85. Daniel "Man Hunting in Malaya," 257.

86. The author searched every issue of *The Royal Journal of Artillery* from 1948 to 1970, available in bound copies at the US Army Combined Arms Research Library. The only articles directly related to the Malaya were the aforementioned "Man Hunting in Malaya," book reviews of *Shoot to Kill*

and *The Long Long War*; and "Presentation of a Silver Centre-piece to the Federation Artillery," a one-page article detailing the activation ceremony for the Federation RA.

87. Smith, *Counter-Insurgency Operations 1*, 68-69; J. P. Ferry, "Full Employment," *Royal Journal of Artillery* 92, No. 1. (March 1965): 21-29; R. G. S. Bidwell, "Gunner Tasks in Southeast Asia," *Royal Journal of Artillery* 91, No. 2 (September 1964): 85-96; C. Beeton, "Locating in Borneo," *Royal Journal of Artillery* 96, no. 1 (March 1969): 14-18.

88. Karl Hack, "The Malayan Emergency as Counter-insurgency Paradigm," *Journal of Strategic Studies* 32, No. 3 (2009): 22. Hack lists several of these aspects and traces their adaptation in other counterinsurgencies within his conclusion.

89. John A. Nagl, *Learning to Eat Soup With a Knife: Counterinsurgency Lessons From Malaya and Vietnam* (Chicago: University of Chicago Press, 2002), 73.

90. Nagl, *Learning to Eat Soup With a Knife, 73*.

91. Larry E. Cable, *Conflict of Myths: The Development of American Counterinsurgency Doctrine and the Vietnam War* (New York: New York University Press, 1986), 90. Cable does not provide a citation for this assertion.

92. Edward E. Hoffer, "Field Artillery Fire Support for Counterinsurgency Operations: Combat Power or Counterproductive?" (Monograph, School for Advanced Military Studies, Fort Leavenworth, KS, 1987), 16-17. This study also reflects the erroneous notion that Templer's hearts-and-minds approach obviated the need for firepower.

93. Barber, *The War of the Running Dogs*, 111.

94. Komer, *The Malayan Emergency in Retrospect*, 26.

95. Miers, *Shoot to Kill*, 39.

96. Komer, *The Malayan Emergency in Retrospect*, 26. As noted above, the issues with low funding were eventually overcome by the boost of Malaya's tin and rubber economy during the Korean War.

97. Komer, *The Malayan Emergency in Retrospect*, 27.

98. Komer, *The Malayan Emergency in Retrospect*, 32.

99. Federation of Malaya, 20.

100. Federation of Malaya, 50.

101. Wade Markel, "Draining the Swamp: The British Strategy of Population Control," *Parameters* (Spring 2006): 39-40.

102. Stubbs, "From Search and Destroy to Hearts and Minds," 114-115.

103. After studying the Malayan Emergency for RAND, Komer would go on to head the United States' CORDS program in Vietnam. Both he and the program are discussed at length in Chapter Four.

104. Figures for resettlement are drawn from Stubbs, 107. Figures for regrouping are drawn from Coates, 91.

105. Markel, "Draining the Swamp," 36.

106. Komer, *The Malayan Emergency in Retrospect*, 58, 60. Due to the Malayan climate, rice generally spoiled after 24 hours. For an account of the effect of hunger motivating CTs to surrender and their efforts at illegal cultivation, read Richard Miers' and Frank Kitson's accounts with SEPs in *Shoot to Kill* and *Bunch of Five* respectively.

107. Markel, "Draining the Swamp," 38-39.

108. Richard L. Clutterbuck, *The Long, Long War* (New York: Praeger, 1966), 40-41.

109. Coates, *Suppressing Insurgency*, 89.

110. Komer, *The Malayan Emergency in Retrospect*, 40.

111. For an illustration of this, refer to Kitson's account in *Bunch of Five*, 80-112.

112. Malaya I Panel Discussion, US Army Command and General Staff College Art of War Scholars Seminar, 4 January 2011, Fort Leavenworth, KS.

113. BI070, Retired General Officer.

114. Komer, *The Malayan Emergency in Retrospect*, 52.

115. Smith *Counter-Insurgency Operations 1*, 17-18.

116. Miers, *Shoot to Kill*, 77. This is corroborated by Kitson's accounts in *Bunch of Five* and the enemy analysis in the operations orders for Operation Nassau and Operation Apollo.

117. Gregorian, "Jungle Bashing in Malaya," 344.

118. Miers, *Shoot to Kill*, 150.

119. Barber, *The War of the Running Dogs*, 149. *War of the Running Dogs* provides a narrative one rubber plantation that withstood 25 attacks occurring over a two-week span in summer of 1951. The plantation only had two veterans from Palestine assigned to defend it.

120. The State Secretariat of Central Perak, *Central Perak SWEC Operation Order No. 1-57: Operation "Ginger"* (1957), Annex A.

121. The State Secretariat of Pahang, 4-6.

122. D. J. Wilford, *Some Aspects of Anti-Terrorist Operations: Malaya* (Fort Bragg, NC: 7th Special Forces Group, 1963), 6.

123. Miers, *Shoot to Kill*, 149.

124. Clutterbuck, *The Long, Long War*, 42-44.

125. Refer to Wilford's *Some Aspects of Anti-Terrorist Operations: Malaya* for some anecdotes of mortar use during ambushes late in the Emergency (1958), 16 and 18.

126. Miers, *Shoot to Kill*, 56-72. His chapter "Yellow Pin" describes this operation in detail, the name referring to the colored map pin which marked the target on his command post's mapboard while they planned this mission.

127. *ATOM,* Chapter 18, Section 3.

128. Sunderland, *Army Operations in Malaya, 1947-1960*, 231-232.

129. Forster, *A Comparative Study of the Emergencies in Kenya and Malaya*, 75.

130. *ATOM,* Chapter 18, Section 4.

131. Gregorian, "Jungle Bashing in Malaya," 344-346.

132. Gregorian, "Jungle Bashing in Malaya," 346.

133. *ATOM,* Chapter 18, Section 6.

134. *ATOM*, Chapter 18, Section 4.

135. Sunderland, *Army Operations in Malaya, 1947-1960*, 238.

136. Forster, *A Comparative Study of the Emergencies in Kenya and Malaya*, 64.

137. Forster, *A Comparative Study of the Emergencies in Kenya and Malaya*, 71.

138. *ATOM*, Chapter 18, Section 4.

139. Coates, *Suppressing Insurgency*, 167.

140. Daniel, "Man Hunting in Malaya," 260.

141. Daniel, "Man Hunting in Malaya," 259.

142. Gunner James Lamb's war diary indicates that they were, on occasion, used in support of food-denial operations due to the intense manpower those efforts required.

143. Sunderland, *Army Operations in Malaya, 1947-1960,* 239.

144. Marston, "Lost and Found in the Jungle," 102.

Chapter 4
The American Experience in Vietnam

> In South Viet-Nam, the West is still battling an ideology with technology, and the successful end of that Revolutionary War is neither near nor is its outcome certain.
>
> — Bernard Fall, *Street Without Joy*

> The end of the fight is a tombstone white
> With the name of the late deceased,
> And the epitaph drear: "A Fool lies here
> Who tried to hustle the East."
>
> — Rudyard Kipling, *The Naulahka*

In Vietnam, American artillery units fired an incredible amount of munitions from a network of distributed base areas, which became an important aspect of the war itself.[1] This method of employment did not directly counter the insurgency via pacification or development because artillery units strictly focused on the delivery of indirect fires. This was mainly due to search-and-destroy operations' requirements for tactical fires, and the organization of the counterinsurgency force. The American experience in Vietnam illustrates that even the most capable and adaptive massed fires to defeat enemy main force elements do not address the drivers of instability in an insurgency if they are part of an ill-suited operational approach in a complex environment.

The Vietnam War and the Malayan Emergency share several superficial similarities and provide a seemingly convenient vehicle for comparisons.[2] Both wars are comparable in terms of terrain, the Maoist origin of insurgent methods, grievances arising from a breakdown in local governance, and an intervening force that could not impose direct reforms on the host nation. However, these similarities are outweighed by many glaring contrasts. In Malaya, the insurgency was a subset of a distinct ethnic minority, but in Vietnam there was no major fault along ethnic lines.[3] The MRLA remained scattered and isolated in the jungles, while in Vietnam, the enemy periodically massed in large mobile formations yet maintained a complimentary capability of discrete subversion at the village level. In Malaya the MCP had virtually no external aid or safe havens, but in Vietnam these factors practically defined the strategic context. Finally, Malaya was a sideshow to the Korean War, while Vietnam was a worldwide media focus.[4] Any comparison between the two wars must account for these nuanced differences.

A major difficulty in analyzing Vietnam through the lens of counterinsurgency is that the war itself was a hybrid of several forms of warfare. The American experience in Vietnam was neither a counterinsurgency nor a conventional war. It was an amalgam, with elements of both at various times and places. This hybrid nature is precisely why it was such a difficult conflict to understand contemporaneously, and such a difficult environment in which to develop a well-framed operational approach.[5]

In addition to this hybrid nature, the war also had several layers. There was a layer of conflict along the lines of a competition in governments, with an unstable central democracy contending with a Marxist-Leninist rural model. There was a layer of conflict across social strata, with an emerging urban middle class challenging a traditional agrarian majority. There were even traces of religious conflict, as the Catholic minority gravitated toward the forces of the government.[6]

Furthermore, this conflict blurred the lines between strategy and tactics. Emerging technology and elements of media, mobility, and firepower challenged the existing categories of warfare. One Vietnamese General's analysis is that the Americans and their South Vietnamese partners had sound tactics but employed the wrong strategy, while the North Vietnamese had the wrong tactics but employed a sound strategy.[7] Tactics and strategy both had an important influence on the US military's employment of artillery units in this war.

<u>Background</u>

As with the Malayan Emergency, the terrain and geography had a major impact on operations. The Republic of South Vietnam (abbreviated SVN)[8] was 67,000 square miles of varied terrain, roughly the size of Florida. This included a 1,500-mile coastline on the south and east, and a 950-mile border with Laos and Cambodia to the west. The distance between the border and the sea varies between roughly 30 and 100 miles. The terrain along the coastline is generally open and flat, with much of the border regions containing central highlands with rugged jungle peaks of 3,000 to 8,000 feet of elevation.[9] These central highlands were sparsely populated, with a majority of the rural population density spread across the fertile Mekong River Delta region of the south. Coastal urban centers dominated the central and northern regions of the country.[10]

Demographics reveal some of the divisions within the country at the outset of conflict. Of an estimated population of 15 million, a 1960 census showed tribal minorities in the central highlands totaling 1 million. Of the remaining population, there was only a 15 percent ethnic minority of Khmer (Cambodian) and Chinese.[11] Religiously, there were roughly 12 million self-identified Buddhists and 2 million Catholics, with small minority communities of Cao Dai and Hoa Hao practitioners in the remote regions of the delta.[12]

Historically, Vietnam had French colonial administration and rule since the 19th century. What distinguished Vietnam from most other European dominions across the globe was that the Vietnamese were expanding from their northern base in a southward colonial fashion of their own at the same time European powers began competing in southeast Asia.[13] This provided for a somewhat antiquated but ingrained view among the Vietnamese that there was a significant difference between the northern and southern regions in Vietnam. Nominal French rule continued through the first stage of World War II, since Japan allowed the sustained administration of Vietnam by a government loyal to the Vichy French. This uneasy setup lasted until metropolitan France was liberated by Allied forces, and the French government in Indochina received direct support from American OSS and British Force 136 teams.[14]

Japan acted swiftly and destroyed the French presence in Vietnam through overt force in 1944. America refused to send combat air support in Vietnam, even as British allies made a supreme effort to support the French from airbases from the west. The Japanese won a swift victory. They garrisoned Vietnam with more than 50,000 troops, and simultaneously informed the Vietnamese Emperor Bao Dai that his country was independent on 10 May, 1945.[15] The Viet Minh began as a resistance force to the Japanese occupation, supported by both Chinese forces and the remaining American OSS teams.[16] While the Viet Minh did not succeed in ejecting the Japanese from Vietnam, they did gain considerable structure, experience and popular support. After Japan surrendered to Allied forces in 1945, chaos ensued in Vietnam as the Viet Minh, Chinese forces, British forces, American advisers, and French POWs all sought to establish varying degrees of control and governance.[17] Bernard Fall offers a description of the emerging dynamic in Vietnam as the French re-established colonial administration:

> Viet-Nam, as a unified independent state had again disappeared. . . . So again Viet-Nam was divided into two distinct states, but from 1946 until 1954 this was to be a new, strange urban-rural division rather than a north-south division, with the Viet-Minh holding much of the countryside including the hill-tribe areas, while the French and later the non-Communist Vietnamese administration were to hold the lowlands, and especially the cities.[18]

In 1953, after nearly a decade of counterinsurgency, French airborne units established a lodgment and airfield near Dien Bien Phu in order to establish an operational reach into Laos and interdict key Viet Minh routes. General Vo Nguyen Giap responded by moving an army of more than 50,000 to encircle this position, and leveraged a supreme effort by the population to sustain the siege on Dien Bien Phu.[19] As it became clear that the Viet Minh would prevail unless there was an outside intervention, President Eisenhower and the Joint Chiefs of Staff (JCS) considered options including air support, a coalition of ground forces and even tactical nuclear weapons. In the end, no American military support was offered to the beleaguered French defenders.[20] In the face of constant artillery bombardment and ground attacks, the defenses at Dien Bien Phu fell in May of 1954. French officers and historians correctly view this as a defeat, not a surrender.[21]

As a result of the tactical and operational defeat at Dien Bien Phu, the French could not continue operations without a significant broadening of the strategic framework in Indochina. As of 1954, the United States began to send all military assistance and support directly to the provisional governments in Indochina rather than through the remaining French apparatus. To this end, it established a Military Assistance Advisory Group (MAAG) on the Korean War model of rolling out conventional divisions in "an assembly-line fashion."[22] Communist subversion continued in the south, and by late 1957 it had evolved into outright terrorist attacks on the village mayors and administrators who formed the backbone of rural governance in Indochina. This number steadily increased to the point of more than 4,000 assassinations and killings in 1960 alone.[23] On 21 October 1960 the war took on a new character, as more than 1,000 Communist troops infiltrated into Kontum and crushed the ill-equipped positional army garrisons.[24] In response to this, the JCS activated an expanded advisory mission, the Military Assistance Command–Vietnam (MACV). In 1962, MACV superseded the MAAG to coordinate all security activities within SVN.[25]

The Insurgency

As with any counterinsurgency effort, the insurgent facet of this hybrid war exploited some very authentic grievances among a disaffected population in South Vietnam. In a study of the Long An province, historian and political scientist Jeffrey Race illustrates that the Communists recognized the need to redistribute wealth and income via land reform. Accordingly, their land tax was effectively less than the combination of the government's land tax and additional rents. The Communists also understood the need to redistribute power and wealth because preferred jobs required the equivalent of a college degree and therefore precluded the rural lower and middle classes. Communist-structured jobs at the village and district-level provided the poor with upward mobility and increased status, albeit with a single avenue for advancement.[26] With this focus, early subversion deliberately ignored the smaller urban population, choosing to focus on the vulnerable rural population of more than 10 million. They reinforced this with a claim that President Ngo Dinh Diem's regime of urban elites had taken over two-thirds of the agricultural land in South Vietnam, further deepening an existing wedge between the urban and rural populations.[27]

These insurgent methods exposed a critical vulnerability of the GVN. During the years of French colonial administration, ethnic French citizens took not only the middle class but also the lower class government jobs. This left an inordinate gap of bureaucrats in Diem's emerging government of Vietnam (GVN).[28] Diem struggled to find an appropriate political theme, shown by his statement in 1956 that "democracy is neither material happiness nor the supremacy of numbers . . . [but] is essentially a permanent effort to find the right political means of assuring all citizens the right of free development."[29]

In his 1972 debrief, Major General John H. Cushman illustrated the difficulty of the American military to understand the Communists' organization and methods, lamenting that "after all this time, [the enemy] is still far from understood and capable of surprise."[30] One of their main advantages was a pragmatic organization which capitalized on the benefits of seeing the revolutionary movement in terms of a social process instead of a military process.[31] Communists sought to achieve concepts such as the balance of force, security, and victory, but these were seen through the lens of social power ratios instead of military power ratios.[32] This reflects the fact that in Giap's model of warfare, propaganda was almost more important than military strength, although the two should be inseparable in accordance with the application of Lenin's theories of revolution.[33] They were also decentralized instead of district-minded, which allowed them to raise more forces through attractive policies at the local level and establish critical links between the people and the Party.[34] In this manner the Communists were assimilated forces, which ensured that they were local and representative, performed functions directly useful to the local community, and were both incentivized and resourced by the local community. As a result, they did not need to protect lines of communication or critical infrastructure.[35]

Communist methods complemented their organizational structure, and were also adaptive to the requirements of local subversion. An analysis of these methods and tactics yields critical insight to the utility in creating a hybrid war of both regular and irregular components. Their techniques and practices were potent. After decades of mixing conventional arms with insurgency, Giap referred to it as a weapon in and of itself:

> Our weapon is the invincible people's war, and we have gained experience in conducting it. That is to say that if there is a greater invention than atomic weapons, that is, the people's war, then the Vietnamese people have effectively contributed to the perfecting of this new arm and keeping it firmly in their hands.[36]

This reflects the fact that Giap's aim was to keep US and GVN forces focused on large-scale operations to address the conventional threat and keep them away from successfully pacifying the coastal areas which the Communists prized most.[37]

In this manner, the dual means of conventional war and insurgency exhibited a symbiotic relationship. Conventional means gave insurgent means ample breathing room, while insurgent means gave conventional means ample support. Through 1966, the Communists' insurgent military force, the Viet Cong (VC), required only 12 tons of supply per day from the north via the infiltration networks in Laos and Cambodia.[38] Besides the most technical weapons and highly-trained specialists, all other requirements were self-sustaining, a phenomena that was exposed as early as 1964 but discounted by MACV.[39] To coordinate efforts across all fronts, North Vietnam established the Central Office for South Viet Nam (COSVN) which harmonized operations for subversion, terrorism, and open military operations within SVN.[40]

Of special interest to this study, the VC and the North Vietnamese Army (NVA) relied primarily on mortars and rockets as their artillery arms. Larger artillery pieces were not regularly moved into South Vietnam until conventional incursions later in the conflict because of challenges with mobility and logistics over infiltration routes. After the exclusive use of mortars until early 1967, the VC adopted the rocket as the weapon of choice after a successful attack on the American air base at Da Nang. Within six months the 122mm rocket system was the "prime weapon" for indirect fires.[41] The American military and the Army of the Republic of Vietnam (ARVN) fought these weapon systems with aggressive patrolling, force protection, and counterfire measures. A comprehensive guide to VC and NVA artillery equipment and techniques was not promulgated until 1970.[42]

The Counterinsurgency Effort

Although GVN officials sensed an external attempt to overthrow the system of governance, they did not sense the attempt to overthrow the social system itself.[43] The GVN did not have the resources to militarily secure rural areas, and soon they realized that they were past a critical point at which the government could not provide enough security to the local populace.[44] MACV saw these early efforts in terms of a military enterprise as well, which is somewhat reasonable given that they were a military organization tasked with supporting a government that saw the issue as a security problem. However, this viewpoint extended to American political leaders and national strategists who eventually adopted a similar military-centric outlook on objectives in Vietnam.[45]

The first attempt to address security and pacification in rural areas took the form of the Strategic Hamlet program. This effort to extend governance to the countryside and provide a local counter-organization to the VC had theoretical roots in the Malayan Emergency, and the British imprint on this program is evident in the way it was carefully planned to be gradually executed.[46] Unfortunately, the execution and scope of the Strategic Hamlet program left much to be desired, as it did not garner early support from MACV, did not incorporate enough local security, and began hastily in the richer delta region due to political connections with the Diem regime.[47] MACV's advisers viewed this as an attempt by the GVN to extend governance into the countryside, while the VC viewed this as an opportunity to insert their "sleeper cells" into the villages themselves.[48] In any case, the eventual collapse of the Diem regime effectively terminated the program.[49]

The Strategic Hamlet program failed to achieve any cohesive *strategic* effect, and by 1963 there were an estimated 23 VC battalions operating in the Mekong Delta, the very region where the program was initiated.[50] In the face of these challenges, the American military sought to disrupt VC infiltration routes into South Vietnam rather than address their support structure within the countryside itself. In May 1963, the JCS directed the Commander-in-Chief, Pacific to develop a program for SVN hit-and-run operations in North Vietnam, which were to be nonattributable and carried out with American equipment, training, and advisory assistance. The plan was approved in September 1963, without being forwarded to the White House for approval.[51]

Measures such as this demonstrate the willingness and ability of MACV and the larger US military to address operational and strategic issues on their own, apart from governmental oversight in the GVN. Diem's regime was increasingly nepotistic and insular, and in November 1963 he was assassinated during a *coup d'état* in which the United States deliberately chose not to intervene.[52] Diem was replaced by a short-lived military junta, which was replaced in another coup by General Khanh in January 1964. In standing by during these turbulent political machinations, American leaders felt complicit in the events, and therefore obligated to the GVN's survival rather than a single, unifying leader.[53]

MACV's operational scope steadily expanded, and in January 1964 General William C. Westmoreland arrived to lead the buildup of advisers, logistical assets and headquarters elements already totaling more than 16,000 personnel.[54] Ambassador Maxwell Taylor and Westmoreland enjoyed a prior relationship that translated into the MACV Commander serving as a deputy to the ambassador for military affairs; this was not an official policy directive.[55] As part of his orientation to the theater of operations, Westmoreland and others visited Malaya for a tour with Thompson in 1964 to speak with participants in the Malayan Emergency and glean lessons from that conflict.[56] Westmoreland visualized the conventional component of the effort and the clearance of VC units in the countryside as a war of attrition. This was due to the fact that, in Westmoreland's view, the political and strategic limits placed on the counterinsurgency effort in Vietnam left no other alternative.[57]

To provide security for the pacification effort at a local level, the territorial forces which were originally conceived in 1961 began to materialize as Regional Forces (RFs) and Popular Forces (PFs) in 1964.[58] This gave the GVN a force to fight an insurgency that had grown from a "brush fire subversion," since they had to focus the conventional forces of ARVN along the border.[59] RFs constituted a military force at the disposal of a district-level or provincial-level leader, while the PFs served a military function for local security in individual villages. Westmoreland accepted the fact that these forces were being created more in the image of an American military force than a force to mimic effective insurgent capabilities, or a even paramilitary force comparable to the Home Guard in Malaya. After the war he argued:

> What was the alternative – small mobile forces not unlike guerrilla units? By very definition the insurgents have the initiative, capable of striking where, when and in such strength as they wish and are able to muster, unless government forces can intercept them in advance. . . . Except for some special units, government forces, unlike guerrillas, cannot be elusive.[60]

This quote is illuminating since Westmoreland's conventional understanding of concepts such as "initiative" and the ability to "intercept them in advance" would govern the tactics of search-and-destroy operations in upcoming campaigns.

The Gulf of Tonkin incident in August 1964 served to increase the international tensions regarding the war, but it did little to expand the scope of the war itself.[61] Aside from the implications of the Presidential use of military force prior to legislative branch approval, the strategic implication for Vietnam was that President Johnson only declared a limited response toward North Vietnam. In his press conference, he stated that "we still seek no wider war," though the response eventually included bombing selected targets in North Vietnam itself.[62]

The lack of an adequate and effective strategy through 1964 prompted discussion regarding the deployment of American ground forces in an offensive role rather than strictly an advisory and supporting role. In 1964, General S. L. A. Marshall[63] saw two courses of action: the counterinsurgency effort could be transitioned to GVN-only capacities and the Americans withdraw immediately; or a large deployment of troops with "enough tactical force to bring off a decision" on the part of North Vietnam.[64] The Americans could only gather limited support from a coalition that never included European allies with major military or civic action assets.[65] As 1965 approached, Westmoreland and his MACV staff realized that the disjointed strategy of defending large bases to be used for bombing North Vietnam's limited military hard targets was having little effect in SVN.[66] The overall goal for the counterinsurgency effort at the time was to provide breathing space for the GVN to consolidate power and stand up for itself, but the prevailing perspective was that the government, and society in South Vietnam, was about to collapse without intervention. Intelligence and local leaders' intuition indicated that in some regions the Communists were making the transition from a Maoist second phase of guerrilla warfare to the third phase of open, mobile warfare.[67] The national government in SVN was in chaos, so no pacification programs were coordinated or effective enough to counter the hyper-organized Communist threat. One American political-military analyst said that even he could never quite tell who was in charge within the GVN during these years.[68]

The ensuing discussions on a refined strategy in Vietnam were to set the framework for a significant part of the buildup and escalation in SVN. Aides close to the President at the time viewed these discussions as having three camps:

1. Secured air bases for the continued but limited bombing of North Vietnam, favored by the Johnson administration.

2. A strategy of linked defended enclaves, favored by Ambassador Taylor.

3. Tactical search-and-destroy operations, favored by MACV and the JCS.[69]

The strategy of defended enclaves was the initial justification for deploying American combat troops to Vietnam, with the goal of securing base areas for ARVN to attack the VC and NVA forces in the countryside. Two combat divisions with 17 battalions total deployed to defend these enclaves, which Taylor wanted restricted to the coastal areas in order to minimize direct contact by American forces and keep casualties low.[70]

Westmoreland had another long-term vision for South Vietnam once the initial force secured these enclave areas and emboldened the GVN. His plan for 68 battalions would initially secure a network of logistics bases and conduct occasional counters to large-scale enemy operations in order to "halt the losing trend." In the second phase of operations, this force would gain the initiative via attacking the VC and NVA sanctuaries, and begin to reinstitute pacification programs. The third phase would consolidate these gains and push the enemy back across the Cambodian and Laotian borders to be contained by an improved ARVN. Pacification and expanded assistance to ARVN and territorial forces would run throughout the offensive.[71] On 28 July 1965, President Johnson announced the commitment of a force package with an additional 44 battalions, with the important caveats that Westmoreland sought: additional troops as needed and the freedom to maneuver.[72]

By November 1965, American forces were beginning to gain and maintain contact with both VC and NVA forces in search-and-destroy operations. Lieutenant Colonel Harold Moore, commander of some of the first airmobile troops to see combat in the Ia Drang Valley, cited the need to establish a working group to study the strategy and get the GVN to reestablish control in the areas cleared by combat forces:

> If they couldn't make it work in Bong Son – where the most powerful American division available had cleared the en-

emy forces from the countryside – how could they possibly hope to re-establish South Vietnamese control in other regions where American military presence was much weaker?[73]

This shows the emerging trend facing MACV in 1966: how to capitalize on the ability to clear the enemy in the face of an institutional inability to hold or develop those areas. As 1966 progressed, more large-scale search-and-destroy operations successfully kept enemy main force units away from population centers, but they did not focus on integrating pacification efforts.[74] Units tried to incorporate civic development where they could, with the US Marine Corps (USMC) units around Da Nang making an effort to distribute Combined Action Platoons (CAPs) to the village level. CAPs were focused to destroy the insurgent support network in the villages, protect the population, organize local intelligence nets, and train their partnered PF platoons.[75] CAPs were independently successful with a lower casualty rate than their counterparts conducting search-and-destroy missions, but USMC leaders failed to link their efforts operationally. Without a unifying theme, and with Army leaders casting this as a "do-nothing" approach, the CAPs technique did not gain a theater-wide implementation.[76]

Search-and-destroy operations continued through 1966 with increasing scope and intensity, especially in the northern provinces.[77] MACV drafted plans to more than double the American troop levels, taking them from 184,300 to 429,000 in an attempt to reach the elusive crossover point in attrition warfare where the losses inflicted on the VC and NVA would exceed their ability to replace those casualties.[78] Unfortunately, the crossover point was almost impossible to reach within the framework of a politically limited war. By limiting MACV's operational and theater strategic options to counter the VC and NVA, President Johnson unwittingly gave Giap the ability to transfer 60,000 troops from guarding the De-Militarized Zone (DMZ) against incursions.[79] Thompson viewed the pursuit of a crossover point in the midst of strategic limits as a farce, noting that "all the people of North Vietnam had to do between 1965 and 1968 was to exist and breed."[80] Even in the midst of continuing tactical success by American forces, it became clear that a new approach must be considered. Unfortunately, most discussions in 1966 and 1967 were framed with pacification seen as a corollary to conventional operations.[81]

One exception was a March 1966 report titled "PROVN" which identified an operational approach of pacification along security, political, and economic lines.[82] The report contended that strategic victory could only be achieved if the rural population willingly supported the GVN individually. Therefore, the actions at the village, district, and provincial levels required a greater emphasis; the war should be fought there and not just to deny the VC and NVA sanctuary and infiltration.[83] PROVN contained key recommendations that addressed the lack of a unified effort in Vietnam:

1. Concentrate operations at the provincial level.

2. Give rural construction primacy among joint MACV and ARVN efforts.

3. Authorize direct involvement of US officials in local GVN affairs.

4. Designate the US Ambassador as the sole manager of all US activities.

5. Direct the sole manager to develop a single, unified plan.

6. Re-affirm the goal of a free and independent non-communist South Vietnam.[84]

Although the report focused mainly on pacification and organizational changes, it was not a soft-handed approach or an attempt at appeasement; the first of four specific initiatives PROVN recommended was to "destroy PAVN and Main Force VC units and base areas."[85] To balance these requirements, PROVN advocated a refined organizational approach:

> Social change within SVN, at this point in time, hinges on US provision of an institutional scaffolding capable of assisting, and, if necessary, assuming a primary agent role in the introduction of needed change. The presence of such an effort, coupled with the already existent military scaffolding, could provide motivation consistent with the degree of pressure required to produce change.[86]

Westmoreland mistakenly believed that many of the actions PROVN promoted were already being pursued, and the unity of effort between MACV and the GVN was eschewed to avoid making the regime look like a puppet government.[87] The team behind PROVN was not alone in its assessments, and Secretary of Defense Robert McNamara's visit to SVN in October 1966 left him with the impression that pacification and a long-term shift in policy were in order.[88] These notions were integrated into the strategic framework to a certain extent, but the campaign plan for 1967 only cast pacification as an ARVN task, with the American forces still firmly rooted in search-and-destroy missions to provide security for that effort.[89] Ironically, Giap may have been the only one to publicly expose this inadequacy, writing in an open source that: "Precisely this makes it impossible for the aggressors, though they may bring in several hundred thousand men, strategically to avoid losing command of the situation and compels them to scatter their forces defensively as well as offensively."[90]

The call to increase support for pacification did not completely fall on deaf ears, and in February 1967 the Office of Civil Operations (OCO) was formed under Deputy Ambassador William J. Porter. The organization had six departments, drawing personnel from the United States Agency for International Development (USAID) and the Central Intelligence Agency (CIA): Refugee, Psychological Operations, New Life Development, Revolutionary Development Cadre, *Cheu Hoi* (reconciliation of surrendered Communists), and Public Safety.[91] The OCO had parallel organizations in each province, and furnished liaisons to major ARVN commands due to their priority for pacification resources and efforts.[92] Some within the MACV staff felt that the OCO was the Johnson administration's ploy to gain US primacy in the pacification mission, with an eventual transition to MACV itself.[93] Because individual agencies maintained control over their assets, the OCO functioned primarily as an instrument that attempted to harmonize pacification efforts, but could not direct them.[94]

The two largest search-and-destroy operations occurred early in 1967 with the multi-divisional Operation CEDAR FALLS clearing the Iron Triangle area in January and the multidivisional Operation JUNCTION CITY clearing War Zone C in February.[95] Following these operations, many American critics within the Johnson administration and SVN saw evidence that the continued use of search-and-destroy tactics were not working. CIA reports showed that the VC and NVA were primarily supplied from within SVN, evidence that they could count on vast and loyal resources.[96] Other reporting showed that the VC regained sufficient strength after the introduction of American troops, and that they were increasingly returning to a Maoist phase-two guerrilla warfare method. Additionally, they retained the tactical initiative, initiating 88 percent of all contacts in SVN.[97] Still, 86 percent of all operations by American forces were search-and-destroy missions through September.[98] During a stateside trip, Westmoreland admitted to President Johnson that the search-and-destroy effort could go on forever without a change in policy, either a change in tactics on his end or a change in strategic limits on the President's behalf.[99]

The year 1967 did have a major boost for the counterinsurgency effort, as General Nguyen Van Thieu won open elections in September with over 81 percent of the population voting in the face of VC intimidation. As such, the population saw this as the first legitimate iteration of the GVN.[100] In this political environment of hope, the American counterinsurgency effort made several major changes to organizational and leadership positions. Originally, President Johnson considered appointing Westmoreland as the next ambassador, to which most of his advisers objected. Westmoreland stayed on to head MACV, while Johnson selected his Vice Chief of Staff of the Army, General Creighton Abrams, to serve as a deputy to improve ARVN. He also selected his interim National Security Advisor, Robert Komer, to serve as the deputy for pacification. In this arrangement, the civilian Komer was to take charge of all pacification operations, but not as a deputy commander in the military chain of command.[101] Together, Westmoreland and Komer decided to merge the remaining OCO structure with MACV's Revolutionary Support Directorate. This resulted in the first organization in the counterinsurgency effort which truly directed pacification efforts, the Civil Operations and Revolutionary Development Support program (CORDS).[102]

CORDS did not run through unit chains of command below the Corps level, but instead it ran through GVN administrative divisions down to the district level. This was deliberately structured in order to prevent subordination of pacification efforts by military leaders, witnessed in Operation CEDAR FALLS.[103] CORDS kept the same six departments as OCO and added four administrative departments. With the merging of civilian and military agencies to form CORDS, the new organization had many civilian supervisors for military subordinates and vice versa.[104] The ARVN territorial forces represented the overlap in the CORDS and military spheres of influence, so Komer focused CORDS on their overall effectiveness as a security force while Abrams' military advisers managed their technical and tactical and training via ARVN.[105] CORDS also sought to develop an awareness of the conditions at local levels, and developed the Territorial Forces Evaluation System and the Hamlet Evaluation System. Both programs sought to rate those elements in a linear system, often forcing the evaluator to assign quantitative ranks for qualitative measures. Many of the CORDS advisers who were actually in the field disparaged the correlation of CORDS data and reports to the effectiveness of pacification and *Vietnamization* in their areas.[106] Although the ratings could not be changed by a higher headquarters to ensure accurate reporting, the system was not well received because of its scope and inflexibility.[107]

In November 1967, President Johnson authorized Westmoreland's request for the "minimal essential" force package, rather than the one that Westmoreland saw as the optimum force, since the larger package would require a call-up of American reserve units.[108] This forced MACV to examine the role and effectiveness of the territorial forces more closely. Security forces were conceptually arrayed in three concentric rings for the counterinsurgency effort: US forces and ARVN to fight enemy units massed in battalion strength or larger, RF units to secure the cities and villages from smaller enemy units, and PF units and the local police to counter communist subversion and penetration in the villages themselves.[109] The benefits of these locally recruited forces was easily appreciated, as seen in the *RF-PF Handbook for Advisors*: "They know which families have relatives fighting on the side of the VC. They are recipients of information regarding VC movement, meetings, supplies, and future operations. In most cases, they have grown up in their own operational areas."[110] The territorial force's chief weakness was a lack of support, and a US force's presence remained the biggest correlation to security at local levels.[111] This aspect would be recognized under the most extreme measures in the months to come.

The Tet Offensive of 1968 commenced across SVN on the night of 30-31 January 1968 in the midst of the largest holiday on the Vietnamese calendar, and it culminated within two weeks except in Hue and Khe

Sanh. Crucially, artillery and aviation assets remained on alert during the holiday's planned cease-fire hours due to prior indicators of enemy activity.[112] The VC suffered enormous casualties, and this reinforced support for Westmoreland's approach since the Army was killing the enemy in large numbers. Leaders puzzled over how the enemy could mass in such large numbers with such a depth of penetration, not understanding the breadth of the resources available to them within SVN.[113] Immediately, differing views of the Tet Offensive emerged. Marshall saw the offensive as desperation on the part of the North Vietnamese, and it had the effect of bolstering ARVN.[114] Westmoreland saw it as an attempt to create another Dien Bien Phu (specifically, at Khe Sanh) to prompt a negotiated settlement, and to capture a few northern provinces since the communists never negotiate from a position of weakness.[115]

Besides Hue and Khe Sanh, virtually all vital population centers and infrastructure was regained in short order by ARVN, as US forces isolated the enemy main forces to prevent reinforcement. ARVN cleared Kontum City, Phu Loc, My Tho, and Ben Tre, with only Ben Tre requiring a significant intervention by US forces.[116] Since the A Shau valley had been lost prior to Tet, the NVA was able to infiltrate over a division's worth of combat forces almost directly into Hue from base areas in the highlands. This was the only instance where a NVA unit succeeded in planting their flag atop a major Tet Offensive objective. With Hue's psychological importance as the central city that historically tied the northern and southern regions of Vietnam together, the battle to recapture the city was contested for 25 days. ARVN committed 11 battalions backed by three USMC battalions to clear the city, while two brigades from the American 1st Cavalry Division isolated Hue.[117] At Khe Sanh, American forces faced a determined attempt by Giap to create another Dien Bien Phu at this isolated outpost. Reportedly, it was so important to Giap that he personally visited the area around Khe Sanh himself prior to the operation.[118] Khe Sanh was the key to interdicting NVA infiltration routes from Laos, and it also provided the western "anchor" along the DMZ's defense in depth. Contemporary critiques were that the American forces got tied down at Khe Sanh when the physical terrain of the region held no value. However, it should be noted that the 6,000 defenders at Khe Sanh leveraged an inordinate amount of indirect firepower to attrit and fight off 15,000-20,000 enemy troops. Westmoreland's account of the campaign begs the question of who was tying down whom at Khe Sanh.[119]

The aftermath of the Tet Offensive would shape the remainder of the war in Vietnam. Short-term counterinsurgent reactions included the rearming of territorial forces, additional American troop deployments, and accelerated pacification measures. The long-term effect was a realization that the GVN's military and security forces must stand on their own intrinsic worth and resources, hence a program for the *Vietnamization* of the force. The Tet Offensive was an undeniable loss for the North Vietnam tactically, but a strategic victory for Giap because it set the counterinsurgency effort on the misguided path of long-term *Vietnamization* of ARVN instead of a crushing consolidation and counter-offensive.[120] Although Westmoreland may be maligned for thinking in terms of a conventional linear battlefield, there is merit to his assertion that President Johnson "ignored the maxim that when the enemy is hurting, you don't diminish the pressure, you increase it."[121]

The Tet Offensive showed that the territorial forces must be armed to meet the threat of the enemy's periodically massed main-force threat. A total of 477 outposts were abandoned during Tet in contrast to the relative level of resolve shown by ARVN units.[122] It was clear after Tet that the RFs and PFs were out-gunned by the VC. In conjunction with the first programs for *Vietnamization*, Territorial forces were armed with first-rate American weapons such as the M-16 assault rifle, M-79 grenade launcher, M-60 machine gun, and the Light Anti-tank Weapon.[123] Territorial forces would take on an increased role because the enemy's casualties during Tet allowed counterinsurgent forces to break into smaller formations and consolidate security gains.[124]

Another short-term result of the Tet Offensive was a final deployment of additional US troops. Since the intense fighting placed a new media spotlight on Vietnam, this decision took on a more political character than any previous decision for an increase of military end strength. President Johnson directed his new Secretary of Defense, Clark Clifford, to engage the senior military leadership on this subject. When he pressed the JCS and MACV leaders to articulate an endstate or describe what constituted "victory" in 1968, neither could give him a concrete answer.[125] But even in the face of this uncertainty of the role for additional troops, Clifford did not directly oppose the attrition strategy when an additional 100,000 reservists were called up for this force package.[126]

A third short-term result of the Tet Offensive was the hastening of pacification programs. This is evidence that after Tet, both MACV and the GVN began to see the value in a more holistic approach to the counterinsurgency effort. In an attempt to maintain the initiative, President Thieu instituted a general mobilization along with the Accelerated Pacification Campaign (APC), which was an acceleration of resources rather than a fresh set of programs.[127] Many leaders in both governments had a cautious optimism following the Tet Offensive, and the APC embodied the only method that met the communist threat in the villages head-on and simultaneously supported American policies and goals.[128] Even though the APC was basically a coercive method to establish governance through security at the village level instead of an attractive political policy, CORDS saw it as the most successful program launched by the GVN in the war.[129] An examination of the internal statistics regarding village sympathies reveals this successful trend: Immediately after Tet, less than two-thirds of villages in SVN had effective GVN security, whereas in under two years only 4.5 percent of the villages remained "contested" and just 2.8 percent were under outright "VC control."[130] In another measure of effectiveness for the APC, the North Vietnamese were forced to rely heavily on the Ho Chi Minh trail for resources instead of relying on a sympathetic rural infrastructure within SVN.[131] The APC also improved ties between Americans and GVN administrators at the provincial and district levels, though a sort of distrust remained between American soldiers and their ARVN counterparts.[132] However, one criticism of the APC is that it masked the fact that the GVN was not stronger, and success during the APC was due to the severe weakness of the VC after the Tet Offensive.[133]

This was one symptom that drove the need for a long-term solution after Tet, and the Johnson administration pursued another limited strategy instead of one that widened the war. The concept of *Vietnamization*, or increasing Vietnamese primacy in all areas of military security and governance gained political cachet, was accelerated under the Nixon administration starting in 1969.[134] Upon taking command of MACV in 1969, Abrams saw *Vietnamization* in three phases: transitioning ground combat to Vietnamese forces, increasing their capabilities to defend their borders, and then finally diminishing US presence and assuming an advisory role.[135]

Under the conceptual umbrella of *Vietnamization*, Abrams had an opportunity to fuse the approaches of American attrition warfare and GVN pacification efforts. Although the yearly campaign plan was a Vietnamese document since 1967, the 1969 Combined Campaign Plan was the first one in which the GVN had the objective of protecting the population from Communist control rather than the military destruction of the enemy.[136] Thus, this was the first year in which the American military's focus shifted from search-and-destroy tactics to security for pacification support and training ARVN under the concept of a "One War Plan."[137] As American combat forces began their phased withdrawal in support of *Vietnamization*, CORDS took on an increasingly critical role in harmonizing the effort in SVN.[138]

Combat operations stressed the *Vietnamization* project as soon as it began, with the North Vietnamese launching an offensive in 1969 to regain areas that the VC had lost during the reconsolidation after the Tet Offensive. The partnered American, ARVN, and territorial forces were so successful that some units even used the "tethered goat" tactic to lure enemy forces into engagement areas.[139] After a swift reconsolidation, American and ARVN forces were able to break down into small units to support pacification efforts again, as they had in the post-Tet APC.[140] VC losses were so severe that some units had their resources cut off almost to the point of physical starvation, and many units were forced to return to a Maoist phase-one effort of subversion to reassert themselves in the newly defended villages.[141] American forces were not without internal issues though, as the force began to exert serious discipline issues once the redeployment of forces began in earnest. Westmoreland attributed this to an organizational idleness in many combat units as tactical requirements passed to ARVN, reflecting that "idleness is the handmaiden of discontent."[142]

Vietnamization continued at an accelerated pace, and the results were becoming evident to the MACV staff. Abrams noticed a marked increase in the professionalism and ability with ARVN.[143] By June 1970 there were 31 RF battalions and 232 RF companies in SVN, allowing them to form some mobile groups at the discretion of district and provincial leaders.[144] CORDS significantly reduced their presence in secured areas through 1971 as part of the overall American drawdown.[145]

Ironically, the strategic limits of the Johnson administration were replaced with a pragmatic context that enabled large-scale incursions into NVA safe havens. American-led operations into Cambodia were moderately successful, while the ARVN-led Operation LAM SONG 719 was unable to operationally sever the Ho Chi Minh Trail in Laos.[146] By late 1972, American military leaders knew intuitively that North Vietnam was preparing for a large-scale campaign in SVN, but conventional wisdom at the time was that if pacification was completed, then the enemy could not survive.[147] The North Vietnamese launched a massive conventional offensive on 30 March 1972 (dubbed the "Easter Offensive") which dissipated in the face of American-backed forces. Although this offensive represented a climax in both military and diplomatic efforts, most US Army battalions still in SVN did not directly enter this combat. With American firepower backing them, ARVN and the territorial forces held just enough during the Easter Offensive that the North Vietnamese halted the offensive.[148]

As MACV continued to draw down forces, diplomats brokered a cease-fire in Paris for 27 January 1973, effectively ending American involvement in Vietnam. This caught many military and civilian organizations in a race to hand over responsibilities to ARVN and the GVN to redeploy because progress in the talks was kept close to President Nixon. The State and Defense Departments, the JCS, and the CIA were not even informed of the talks until 1972.[149] President Thieu was in a sort of race as well. Since the GVN controlled 75 percent of the area and 85 percent of the population in SVN, they sought to consolidate gains while they still had the tail-end of American military and civil assistance.[150] Redeployment timelines were accelerated from the initial projection of 69,000 Americans for the 1973 campaign, and the last military forces departed SVN in March.[151] Concurrently, CORDS suspended all operations to allow for a timely withdrawal in 1973.[152] The territorial forces had already displayed their mettle during the Easter Offensive, leading some advisers to conclude that CORDS' work was complete in many areas anyway.[153]

As a postscript to the American experience in Vietnam, the Communists overwhelmed forces in SVN in 1975 with a massive conventional invasion. Giap watched the withdrawal of American forces and the institutional adjustments in SVN attentively, realizing that he did not have to win the revolutionary war in the countryside anymore. Reflecting after the war, he recalls that he needed just enough guerrilla forces to attack and hold GVN weak points so that their limited NVA regular force units could concentrate on specific objectives.[154] The NVA expected to gain key areas in their 1975 offensive to shape the battlefield for a decisive campaign in 1976, and were surprised by their initial success.[155] Since ARVN developed a dependence on American military advisers and firepower, it could not lead or employ fires during the 1975 conventional war as it had in 1973.[156] In retrospect, *Vietnamization* can be considered a success or failure depending on the length of a timeline for the analysis. When measured in terms of President Nixon's goals with an end date of 1973, *Vietnamization* allowed the United States to disengage at a strategic stalemate, with the GVN intact and building civil capacity throughout the country. When measured in long-term results with an end date of 1975, *Vietnamization* failed since it improved ARVN material but did not improve the leadership enough to defend the borders alone.

In any case, the result at the end of the second war in Indochina was a unified Vietnam, firmly under the banner of Communism.

The Employment of Artillery Units

Westmoreland's widespread employment of American artillery units in an indirect fires role to support counterinsurgency efforts in SVN should not come as a surprise because he was an artillery officer himself. His first meeting with Taylor was an effort to sell his battalion's firepower to Taylor and General Gavin during the World War II campaign in Italy while he was supporting the 82d Airborne Division. His perceived influence of artillery was reinforced during a discussion with retired General Douglas MacArthur, the last man he visited before departing for command in Vietnam. Westmoreland's recollection of MacArthur's parting advice was:

> [D]on't overlook the possibility, he concluded, that in order to defeat the guerilla, you may have to resort to a scorched earth policy. He also urged me to make sure I always had plenty of artillery, for the Oriental [*sic*], he said, "greatly fears artillery."[157]

This approach has roots in an institutional American proclivity to exchange firepower and resources for decreased casualties in warfare. Some critics contend that this tendency to shape the battlefield for combat with superior conventional firepower was a driving force in the adoption of attrition warfare.[158] This is evidenced in the unofficial motto of the 1st Brigade, 101st Airborne Division (Airmobile) seen in December 1966: "Save Lives, Not Ammunition."[159] This motto was not completely unfounded, given the political and public pressure to minimize American casualties at the time. Westmoreland viewed the role of artillery in terms of "interlocking fire-support bases, [where] artillery units could support each other."[160] This led to an environment where, at the apex of American troop levels in SVN, 61 artillery battalions supported 59 infantry battalions.[161]

The French experience in Indochina provided a lesson for fire support in search-and-destroy operations, which American tacticians duly applied. Unfortunately, this was not a decisive advantage beyond the actual combat engagement of conventional units. In an analysis of the artillery employed at Dien Bien Phu, it becomes obvious that firepower alone will not secure a region unless the support of the population is secured first. Any insurgent force that controls the rural countryside and has the additional advantage of a safe haven can escalate the level of warfare in the region almost at will. The French invested more than 500 artillery pieces and 100,000 troops in the Red River delta region, but the effort was futile since it was by the means of conventional military security.[162] At Dien Bien Phu, the French actually found themselves out-gunned by the Viet Minh, in large part due to the Viet Minh's control of the countryside and this ability to escalate the level of warfare when it benefitted them. In SVN, the American force took careful measures to never find itself out of range of overwhelming firepower.[163]

If the French Indochina campaigns provided a negative example in the employment of artillery units for counterinsurgency, the Malayan Emergency provided a positive example. In addition to the ongoing confrontation in Borneo, there were several contemporary techniques of leveraging fire support from dispersed artillery units in mountainous jungle terrain.[164] Although no direct link to the British lessons are cited, the 25th Infantry Division Artillery conducted grueling trials in 1962. They developed and promulgated techniques that required only ropes, pulleys, and the organic cannon crew to move a 2.5-ton 105mm howitzer across the water and terrain on Oahu that replicated the mountain, jungles, and delta regions of SVN.[165]

The war in SVN became a proving ground for many innovations and equipment refinements in the artillery world, to include forward observer optics, means of target acquisition, and computerized technical fire direction. However, the introduction of a new array of artillery pieces had the greatest impact on the employment of artillery units. The self-propelled 105mm (model M108), 155mm (M109), and 8-inch (M110) all made their debut, as did the towed 105mm (M102), 155mm (M114), 8-inch (M115) and 175mm (M107). These pieces would become the mainstays of American artillery in SVN with their increased range, lethality, ability to traverse, and mobility.[166]

In general, the virtues of coverage and responsiveness governed the employment of artillery units in Vietnam.[167] In 1964, one of the first contemporary reviews of artillery use in operations by an in-country American artillery adviser revealed that "[g]reater than the need to mass batteries and battalions is the need to cover all vital areas with the greatest amount of artillery available."[168] Initial American operations in SVN showed the difficulty in ensuring this coverage. During the first week of battles in Plei Mei and Ia Drang, towed artillery was out of range to support operations, prompting the first use of airmobile-delivered Fire Support Bases (FSBs).[169] As American maneuver forces expanded their role in SVN, artillery units emplaced more FSBs to provide lethal indirect fires in both direct support and general support. Due to the "customer service" attitude of fire supporters, modern radio communications, and the ability to rapidly resupply FSBs via air, artillery coverage was so good that many infantry companies went on extended missions without their weighty 81mm mortars.[170]

To complete the coverage and responsiveness of artillery, several new platforms emerged. To increase the responsive quality for airmobile units, the army introduced Aerial Rocket Artillery (ARA) units. These units consisted of UH-1 helicopters with large 2.75-inch rocket pods attached. A crucial quality that differentiated them from traditional close air support platforms was that they functioned within the artillery organization, and operated on tactical artillery radio nets during combat. The ARA units were extremely responsive because they could reposition as a combat engagement developed and were used efficiently to engage fleeting targets of opportunity.[171] By 1967, ARA pilots began to self-identify as artillerymen.[172] To increase the coverage of artillery units in the delta region, artillerymen established riverine fire bases with limited mobility.[173] After several failed attempts to mount howitzers to standard US Navy pontoons, the LCM-6 landing craft was adopted as a suitable firing platform with some mobility on the waterways. This development gave American forces in the delta region a means of artillery support in an area with very few trafficable roads.[174]

Artillery units were able to provide a wide coverage of responsive fires, but there was a heavy resource cost associated with this capability. Obviously, aerial and naval platforms were expensive means to fill the gaps in artillery coverage and responsiveness. But in the effort to establish more FSBs, the cost in terms of manpower and combat unit requirements began to compound. Battalions often reorganized from three- to four-battery configurations in order to occupy more firing positions. While they were able to cross level existing gun crews, they required a much higher authorization for support personnel.[175] There were also requirements from other combat arms to support this framework. Although there were great improvements in projectiles and techniques for local defense of the FSBs beginning in 1966, it generally required a full infantry company to secure an FSB that supported one or two artillery batteries.[176] Specific munitions such as the Beehive round[177] and the Killer Junior technique[178] were two key elements of improved positional defense. Still, artillery units did not have enough manpower to simultaneously man the gun positions and provide local defense for an FSB, so there remained a large manpower bill for the supported maneuver unit.

The final factor that increased coverage and responsiveness was the ability to fire in a complete circle from the FSB. The two key developments that enabled this were the new ability to fire artillery pieces in a complete 6400-mil circle,[179] and a new method of plotting fire direction charts in all directions from the firing unit. The use of field-expedient pedestal jacks gave 105mm and 155mm pieces the ability to rotate and fire in any direction from the same gun position, and this technique was distributed throughout the force.[180] Techniques were also developed to rotate even the heaviest artillery pieces, the 175mm and 8-inch howitzers.[181] Concurrently, omnidirectional fire direction techniques were developed stateside and promulgated through the *Artillery Trends* professional journal and were adopted in 1967 as the sole method for plotting and determining firing data in US Army artillery.[182] With the increased distribution of artillery units to remote firing locations in smaller elements, fire-direction centers were established at the battery and even the platoon level.[183]

Artillery units were so distributed that many issues in support specific to artillery arose during sustained operations. During the chaotic environment of the Tet Offensive, many headquarters found it more practical to leave the artillery in place at their FSBs, and instead change the command relationships with the unit they supported.[184] Beginning in 1965, Field Force Artillery Headquarters supported each Corps Tactical Zone (CTZ) to manage the requirements of artillery survey, metrological data support and artillery-specific ammunition management.[185]

The qualities of coverage and responsiveness describe the approach for using artillery units in an indirect fires role, and methods for the operational and tactical employment of these units developed throughout the war. When the first artillery units deployed in 1965, there were few sources available for artillerymen to inform their opinions on the use of artillery in counterinsurgency environments. The 1963 edition of US Army Field Manual 31-16, *Counterguerrilla Operations*, echoes the *ATOM* in its laundry list of possible uses for artillery: assisting in the defense of static positions, harassment, flushing guerrillas into ambushes, deception, illumination, blocking escape routes, and inflicting casualties in "tightening the noose" operations.[186] Although American artillery delivered fires to accomplish all of these tasks in Vietnam, few would be as contentious as the use of fires as harassment and interdiction (H&I). They were effective in this role in Malaya because they were used continuously to dispirit an enemy when his demoralization was an integral part of the overall attempt to turn CTs in the jungle. In Malaya, H&I fires complemented food denial and psychological operations successfully. Within SVN, they were an attempt to use firepower as a substitute for additional patrols. By 1966, 65 percent of all fire missions were unobserved H&I fires. Even with an institutional focus on reducing them due to the intense logistic requirements or supporting artillery at distant FSBs, these unobserved H&I fires only fell to 45 percent of the total in 1967.[187] H&I fires were eventually reigned in only by strict regulation; in 1968, a directive banned unobserved H&I fires unless they had a piece of intelligence corroborating the suspected locations.[188]

The integration of artillery in large-scale search-and-destroy operations evolved swiftly, and a comparison between Operation CEDAR FALLS and Operation JUNCTION CITY shows evidence of this rapid development in SVN. Operation CEDAR FALLS was a multi-divisional operation to attrit enemy forces in the Iron Triangle area to clear that insurgent base area. The operation commenced in January 1967 with a poorly coordinated and synchronized preparatory fire plan that disoriented pilots as they discharged maneuver forces at initial landing zones, although the fires were sufficient to suppress the enemy.[189] Continuing through the operation, H&I fires were unobserved, so it was only assumed that they were effectively interdicting the enemy.[190] Command and control was also an issue. Due to the large number of artillery units participating in a fluid maneuver environment, battalions had up to five firing batteries, which was more than they could effectively command and control.[191] Although one brigade after-action review praised the proactive fire control of artillery and air assets, the US 1st Infantry Division had such a difficulty in this task that they had to delegate fire control to their direct support battalions.[192]

Operation JUNCTION CITY, another multi-divisional search-and-destroy operation, followed in February and April 1967. This time, objective area was War Zone C northwest of Saigon. Eighteen artillery battalions participated, an indication of improved ability in command and control artillery units. The initial preparatory fires for Operation JUNCTION CITY used a shorter duration of artillery firepower, which was successful in eliminating opposition at the landing zones without the confusion and disorientation of Operation CEDAR FALLS.[193] The operation had better coordination between air and artillery assets, leveraging *ad hoc* Artillery Warning control centers, and the use of artillery rounds marking targets for close air-support strikes.[194] A Battery, 3rd Battalion, 319th Airborne Field Artillery Regiment conducted the war's only drop zone fire mission, delivering artillery fires after jumping in the 173rd Airborne Brigade's parachute assault. Although the mission was a success and validated the concept of airborne artillery support during operations in SVN, there was not another airborne operation large enough to warrant direct support artillery fires.[195] Air assets were also used in a logistical capacity, as the 1st

Infantry Division Artillery used an air drop to deliver critically needed ammunition to a remote FSB during the operation.[196] That same unit resolved fire control issues from Operation CEDAR FALLS, and even integrated target acquisition assets.[197]

Operation JUNCTION CITY did reveal some limits for the employment of artillery units, and firepower in general, within the counterinsurgency effort. Although fratricide was extremely low in the operation, American forces did suffer casualties attributed to probable errors in range associated with firing large volumes of artillery.[198] After initial challenges in penetrating some of the thicker jungle canopies with artillery rounds, high-angle fires were used to great effect.[199] Although the resupply of artillery units was generally sufficient, the supply chain could not manage artillery ammunition lots, which undoubtedly contributed to a decrease in first-round accuracy.[200]

Artillery tactics and techniques improved steadily with each search-and-destroy operation until the massive campaigns to reconsolidate during the Tet Offensive. Operations at Khe Sanh demonstrated the utility of artillery units providing prodigious amounts of firepower in SVN in a conventional, positional battle. By one measure, American forces expended more than 1,800 tons of munitions per day during operations to defend the base at Khe Sanh.[201] Westmoreland took a lesson from General Vanuxem, a French commander at Dien Bien Phu, that the French had neither sufficient artillery to support them from outside of Dien Bien Phu nor the surrounding high ground for observation. As such, he directed the MACV staff to study the conflict in order to draw out probable NVA tactics at Khe Sanh. At Khe Sanh, American forces had three 105mm batteries, one 155mm battery, and one 4.2-inch mortar battery. But more importantly, they also had sixteen 175mm guns at Camp Carroll, well within range to support Khe Sanh. These guns would showcase the importance of coverage and responsiveness, firing more than 1,500 rounds per day.[202] Air strikes were integrated at a level that had not been seen in SVN, with an average of 45 sorties of B-52 bombers per day from airfields in Guam. When the enemy massed along predictable axes of advance, the Fire Support Coordination Center at Khe Sanh integrated these echeloned air strikes with artillery fires, similar to pre-patterned fire plans from World War I.[203] Illustrating the insurgents' ability to scale back the operational tempo of the battle almost at will, the NVA and VC almost completely disappeared from the region when their losses became too severe and almost all artillery missions returned to the familiar tactics of counterguerrilla and supporting search-and-destroy missions.[204]

After the post-Tet reconsolidation, artillery units began to support pacification operations with increasing frequency. But even in support of pacification, artillery units were to provide indirect fires against fleeting targets instead of expanding the reach of security in a nonstandard role.

Two operations from this phase of the war show the varying degrees of the successful employment of artillery units in support of pacification, Operation WASHINGTON GREEN and Operation RANDOLPH GLEN.

Operation WASHINGTON GREEN was meant to be a showcase for the new approach of support to pacification, and if the operation went well the techniques would be refined and adopted across SVN by Abrams. The 4th Infantry Division Artillery was in support of the 173rd Airborne Brigade, as the maneuver units were split off across the area of operations in platoon- or even squad-sized elements. Units were to defend rural hamlets and improve the qualities of territorial forces; indirect fires were to be strictly limited in order to minimize collateral damage.[205] Artillery was limited to use against rockets and other enemy forces seeking to disrupt pacification efforts, which allowed artillery units to find a balance between providing fires and training classes for ARVN units.[206] The Division Artillery also augmented a civil affairs detachment and directly contributed to the pacification support around their FSBs.[207] However, ARVN artillery units were slow to respond to calls for fire and dangerously inaccurate. When the territorial forces refused to call for ARVN artillery, their advisers acquiesced and employed American artillery fires even though they knew ARVN would not improve without some repetitions. This was a symptom of the larger issue that doomed Operation WASHINGTON GREEN, that ARVN and the GVN simply could not stand on their own yet without significant American support.[208]

Operation RANDOLPH GLEN suffered much the same fate, but with a less focused use of artillery units in support of pacification operations. The 101st Airborne Division (Airmobile) had key tasks to provide security for the population, eliminate the VC infrastructure, complete public projects, improve economic programs, and implement economic reforms.[209] Although the artillery units within the Division Artillery kept as busy as any combat unit, they made no discernable contribution to these goals. Artillery Units within the division made a total of 86 battalion and battery air movements to establish new FSBs. Not only did this use an excessive amount of air transport, but also an excessive amount of infantry companies to assist in local defense of the FSBs.[210] Although medium howitzers were already providing area coverage, light artillery remained in a direct support role to maneuver units. "Confirmed" fire missions accounted for only 79,772 of the 613,864 rounds fired; H&I and Illumination fire missions accounted for even less. A majority of fire missions were registrations, counterfire missions, and suspected targets.[211] Operations comparable to RANDOLPH GLEN stretched the concepts of coverage and responsiveness to its logical limit, since the overall number of artillery units in SVN was drawing down at the time. Units such as the 11th Marine Regiment had distributed firing positions for every asset from 4.2-inch mortars to 175mm guns in 1970, and the larger pieces' long range began to substitute for the decreased number of FSBs.[212]

As the *Vietnamization* effort progressed, the ability of ARVN artillery continued to improve incrementally. At the beginning of the American involvement in SVN, field artillery advisors assigned from MAAG to ARVN units generally consisted of a Major at the Corps and Division level, and a paired Captain and NCO at an artillery battalion. Although they attended a six-week common course at Fort Bragg for all advisers, there was no training specifically for artillery or fire-support advising.[213] MAAG also made an effort to improve ARVN artillery officers' professionalism, and the decade between FY1953 and FY1963 saw 663 Vietnamese officers attend courses at Fort Sill.[214] Advisers had a steep hill to climb, since the French influence on artillery in SVN was undeniable. For years during the French Indochina wars, French observers would send fire commands back to a Vietnamese-manned gun line, mentally calculating adjustments for the gun. This method was crude and prone to errors, but it was also the fastest method available. The consequence was that ARVN units did not have an institutional tradition of fire direction or attention to detail regarding requirements for accurate predicted fire.[215]

Improving ARVN artillery was not a priority until 1969 as *Vietnamization* became a MACV-wide priority. Rapid redeployments of many American artillery units coincided with *Vietnamization*, so it was done within an environment of competing demands to train an allied force and provide indirect fires simultaneously.[216] A prior survey of ARVN artillery from 1963 shows the state of emergency for ARVN artillery at the outset of the American buildup. This study found that ARVN artillery was spread so thinly that even the idle guns at their Field Artillery Training Center were in position and ready to fire. Other units supporting ARVN had similar issues, and one 4.2-inch mortar battalion had 13 platoons supporting a 200km by 60km area. The benefit in this was that fire direction was naturally beginning to migrate to the platoon level, the same phenomenon seen by the British in Malaya and the American partners in SVN. The report concluded that the organization of artillery forces was stressed but valid, and their major weakness was integration with local village and hamlet defensive plans.[217]

At the start of artillery's *Vietnamization,* I Field Force Artillery conducted a four-month study to evaluate ARVN's artillery support.[218] Through initial partnered operations and equipment modernization through 1970, ARVN artillery battalions calibrated more than 88 percent of their pieces by training on American techniques for survey, meteorological data, and registration.[219] As ARVN began to shoulder more of the burden for providing artillery fires in late 1970, advisers noted the same theme seen during Operation WASHINGTON GREEN. ARVN units would still lobby for American air support rather than use their own marginally accurate artillery even if they were in position and ready to fire.[220] There was no

ARVN equivalent for an ARA unit, so the 4th Battalion, 77th Field Artillery (ARA) still flew in direct support of ARVN units. This had mixed results by contemporary American metrics; in these operations it took an average of 16 rockets to result in one enemy kill.[221] By the Spring of 1970, ARVN artillery had improved to the point that they were metaphorically validated during the offensive into Cambodia. All ARVN artillery units supported from static positions during the offensive, which allowed them to concentrate on the basic skills of providing timely and accurate indirect fires. ARVN artillery began the offensive with a coordinated 390-minute preparatory fire plan. They fired more than 200,000 rounds, good for more than one-third of the total fire missions. The chief constraint during this operation proved to be ARVN's own logistic support for their artillery units.[222]

ARVN artillery units proved that they could deliver indirect fires, but still needed much improvement before American forces departed. Special artillery skills, including survey, meteorological support, and ammunition handling, were still extremely poor, in many cases worse than their NVA counterparts.[223] In 1971, one division commander identified this need during his outbrief, stating that ARVN artillery should remain a top priority due to their potential effectiveness, but "that gap must be closed . . . specifically to improve first-round accuracy."[224] Indeed, ARVN artillery had come a long way since being limited to only two rounds per day by Westmoreland in 1966, but there was still room for improvement.[225]

In 1970, the territorial force began improving in artillery integration, even though this was identified as a critical requirement by American advisers in 1964.[226] At the time, RF battalions had a section of two 81mm mortars, and PF platoons had no fire support whatsoever. It was only in 1969 that corps commanders had been instructed to ensure that fire support plans supported the territorial forces in their CTZ.[227] Although advisers had devised a color-coded plan to make indirect fires easily accessible to minimally trained PF forces, ARVN artillery was almost completely unresponsive in some areas.[228] This is evidenced in the records of Military Region Four's records from October 1966 to March 1967, where only 13.2 percent of contacts were reinforced by artillery.[229] Translated field manuals for observed fire techniques were conspicuously absent from a list of manuals to assist village defense plans.[230]

In response to these massive shortfalls, territorial artillery teams of two guns stood up in 1970 to provide fire support in village defense plans.[231] Although there were plans to field actual RF artillery units, the institutional focus on improving ARVN's artillery was clear. By 1971, all but two of ARVN's artillery battalions had been fielded, while the RF had 176 artillery platoons authorized, 100 of which were activated, and only 53 of those activated platoons were deployed in support of operations.[232] It was not until late 1974 that organic artillery units were finally fielded for the RF mobile groups, with a four-howitzer 105mm battery supporting three infantry battalions per mobile group.[233]

The final piece of security forces in SVN were the Civilian Irregular Defense Group (CIDG) units. These elements, advised (and sometimes led by) American Special Forces advisers, did not have an organic fire-support capability. Due to their remote areas of operation in the central highlands to secure the population against VC infiltration, they were not always within range of an American artillery unit. One adviser recalls that he "never got an opportunity to call in artillery fires on the enemy. I dreamed about it, read about it, but never got to use it."[234] In many cases, advisory teams relied on their own abilities with mortars for base defense and the occasional availability of close air support.

Fielding and training the ARVN and RF artillery units was an arduous process, but a necessary one. Difficulties were exacerbated by the continual American withdrawal from SVN and the requirement for responsive fires to support operations. These necessities mostly prevented a one-to-one partnering with ARVN artillery units, but the need was satisfied through advisory and assistance roles. Unfortunately, *Vietnamization* did not begin earlier in the war, for if ARVN had a truly capable artillery force it would have been able to take responsibility for much of the general support requirements, thus freeing a large number of combat troops to serve the counterinsurgency effort in a more pragmatic role. It remains a point of conjecture if they could have contributed much, for the promise of pacification and *Vietnamization* may fairly be critiqued as too little, too late. The following section analyzes the employment of artillery units within this counterinsurgency effort to determine the causes which led them to be engaged in such a manner.

Analysis

With only a cursory overview of the counterinsurgency effort in Vietnam, an initial tendency may be to disparage Westmoreland's approach of attrition and lionize Abrams' approach of pacification support. However, an understanding of the nuances in the earliest operational framework in SVN illuminates several problems with a clear or deliberate division between the two methods. Westmoreland correctly asserts that search-and-destroy operations embodied a tactic, not his strategic approach to counterinsurgency as the media and some analysts portrayed it.[235] He also had an imperfect array of options due to strategic and political limitations; he could affect neither North Vietnam nor the homefront to alter this strategic framework.[236] As with Briggs and Templer in Malaya, there is not so much of a divergence between the preferred counterinsurgency approach between these two generals, but more of a divergence in the organizational structure and political climate each contended with. On several occasions Westmoreland stated that he wanted to unite the pacification effort under MACV, but Taylor kept it as a civilian government function.[237] Finally,

GVN and ARVN leaders themselves attributed the success of the APC effort from 1969 to 1971 to the increased security that Westmoreland's attrition strategy yielded.[238]

This is not to say that Westmoreland's strategy of attrition warfare was particularly well-suited to the environment; it demonstrably was not. But the historical analysis should be framed as one of both adequacy and effectiveness, not as a search for the perfect counterinsurgency in vain. Initial major battles such as the Ia Drang campaign in 1965 gave MACV an illusion of a validated search-and-destroy concept, although there was not a concerted effort to capitalize on these individual successes.[239] Westmoreland assumed that most isolated areas did not hold intrinsic value except when the enemy was massed there. This notion does not address the fact that with temporary search-and-destroy operations as the preferred method, nobody is holding the critical rural and populated areas (which do hold intrinsic value) either since this was before the improvement in territorial forces.[240] Westmoreland himself alluded to this in a message to General Wheeler's forces in August 1965:

In order to be effective we cannot isolate US troops from the population or deploy them solely in jungle areas where they can be bypassed and ignored by the VC. In the long run, we must use them in areas important to the VC and the [government of Vietnam]. . . . The final battle is for the hamlets themselves and this inevitably draws the action toward the people and the places where they live.[241]

By focusing on attrition and an elusive crossover point that could never be reached by blocking infiltration routes, Westmoreland's forces never got into the hamlets of which he speaks. As Thompson critiqued this in 1969, he observed that adding combat resources without integrating the offensive military plan and civil governance development plan was "doubling the effort to square the error."[242]

As a counterinsurgency approach, pacification expressed solely through civic development would not have addressed the lurking enemy that sought to sweep away any gains within the villages. Arguments that a pacified countryside across SVN would have withstood the main force assaults in 1973 and 1975 are disingenuous and unconvincing.[243] Although his approach is more synonymous with that notion, Abrams himself understood the necessity of a large conventional force by acknowledging the idea that "you just can't conduct pacification in the face of an NVA division."[244] Both Westmoreland and Abrams considered pacification, but focused on operations against the main forces in order to keep them away from the population centers.[245] This held more importance to Westmoreland's approach, since the RFs and PFs could not protect the population from enemy main force units during his command.[246] CORDS became effective only after the VC's great conventional defeats of the Tet Offensive, and Komer understood that pacification had always been a priority but not a practice.[247]

The inability of the Americans and the GVN to balance this illustrates the genius inherent in Giap's approach to have different forces or areas in different Maoist phases of insurgency. MACV could never adapt to this, and instead needed an overarching and synchronized (as opposed to a harmonized) campaign plan of one single style. On the other hand, the Communists were able to support both conventional and insurgent operations simultaneously, giving either one primacy depending on the environment.[248] The irony in this was not lost on the CIA's Chief of Station in Saigon, William Colby: "In an ironic asymmetry, the communists initiated a war against Diem in the late 1950s as a people's war and the Americans and the Vietnamese initially responded to it as a conventional military one; in the end the Thieu government was fighting a successful people's war, but lost to a military assault."[249]

The fact that the counterinsurgency effort could not conduct these operations simultaneously is a symptom of a larger issue, that of a disunity in efforts. Throughout the war certain tasks or responsibilities were segregated by organization, such as the ARVN primacy for pacification prior to 1969. Two major inefficiencies of the counterinsurgency effort are exemplified in the lack of unified effort between the American and GVN institutions, and a lack of unified effort within MACV itself.

In the earlier stages of the conflict, the United States and the GVN lacked a unity of effort. The major issue was that regimes prior to Thieu's government fundamentally did not want to address rural power grievances, they wanted to crush the Communist front.[250] Cultural differences certainly played a role as they inevitably will, but some anecdotes from Vietnam border on high comedy. Virtually no American leader spoke a phrase of Vietnamese, so they relied on translators or other elites who could speak English. Since they were almost exclusively from the privileged and Catholic urban minority, and the focus of many operations was the Buddhist agrarian population, it is akin to learning about Iowa farmers from a Harvard law professor.[251] In one case, McNamara wanted to shout "Long Live Vietnam" to a gathering of Saigonese during a visit, but instead roused them with an exhortation that "The Southern Duck Wants to Lie Down!"[252]

The APC served to strengthen the bonds between allies, and is perhaps the high-water mark for these relations. The post-Tet environment in 1968 allowed the US 25th Infantry Division to conduct operations as partnered companies, platoons, and even squads that integrated all manners of GVN assets.[253] However, the APC was not sustained with a follow-up institutional change that may have made these links permanent; it only accelerated resources and coercive measures until the population was secured from Communist subversion.[254]

The process of *Vietnamization* actually served to weaken those same bonds. Many in the GVN and ARVN resented the notion of *Vietnamization* because they had fought longer, and with more casualties, and knew they would continue to fight alone after the American withdrawal.[255] The resulting ARVN force had all of the liabilities of a conventional force which had to conduct a counterinsurgency, but few of the assets. As a result, the damage was more psychological than material when the United States cut off aid prior to the 1975 offensive.[256]

MACV displayed a moderate degree of unified effort, but it also had some shortcomings. Within MACV, Komer identified several contributing factors: military operations were not the repertoire of the other agencies involved, the capabilities did not exist within the armed forces, the lack of a unified GVN–MACV administrative structure, and an institutional focus on conventional operations.[257] With the establishment of CORDS, many of these issues were addressed. By establishing control over their own personnel with an indifference to military and civilian distinctions, CORDS improved on the OCO model and effectively integrated much of its harmonizing function with the MACV staff.[258] CORDS was criticized by some field commanders as just "window dressing" with discrete programs of wells and schools.[259] Even if some of the local-level civic projects were not linked to a wider plan to improve governance, CORDS brought pacification efforts under a single manager and gained military support. This finally provided a unified and direct challenge to the hyper-organized VC structure at the local level.[260]

CORDS also helped to bring the war effort into a framework of legitimacy. By design, CORDS' influence was limited since ultimate orders for pacification came from the GVN. While American leaders always supported the current regime, bureaucratic processes may have masked the intractable nature of social, political, and economic issues within SVN.[261]

This unchanging nature of these social dynamics between the government and the rural population in SVN greatly contributed to the largest shortfall in the early counterinsurgency effort: a failure to isolate the insurgents from their base of support. The other factor was the constantly cited misconception that the guerrilla forces got their support from the North Vietnamese regular forces via the Ho Chi Minh Trail, hence the priority placed on disrupting infiltrations rather than the real base areas. VC requirements from outside SVN were only 12 tons of supply per day in 1966; all else was procured locally.[262] This was always a fallacy within MACV during Westmoreland's command, even though Bernard Fall had identified it as early as 1964 in *Street Without Joy*: "The hard fact is, that, save for a few specialized anti-aircraft and anti-tank weapons and cadre personnel not exceeding perhaps 3,000 to 4,000 a year or less, the VC operation inside South Viet-Nam has become self-sustaining."[263] This phenomenon was continually reported, as in a CIA report from 1966 which

Westmoreland acknowledges in his own autobiography.[264]

An early attempt to isolate insurgents from their base was the Strategic Hamlet program, which ironically would have worked better had it been complemented by the later search-and-destroy operations. Those search-and-destroy operations would be hampered by an inability to separate the insurgent from the population. By using attrition without isolating them, the VC just had to survive amongst the population to win.[265] In 1969, Thompson contended that the earlier Strategic Hamlet program would have failed had it been continued past 1963 since there were more than 4 million refugees within SVN. Even if the population had been resettled in a massive effort by the GVN, it would have only completed the first step. He saw the second critical step as the implementation of population controls, which the Diem regime could not have applied without an effective paramilitary force in 1963.[266] So, at the cost of tearing down the existing intelligence nets within existing communities, the Strategic Hamlet program failed to isolate the VC from their base of support.[267]

The CAP program represented a way to replicate this missing paramilitary capability, in an attempt to isolate the VC from their support networks at the village level. Even though the program showed some initial promise within the USMC units around Da Nang, Westmoreland had the incorrect impression that he would never have enough combat power to support it across SVN, even though a Department of Defense study showed that it would only take 167,000 troops to secure every hamlet in the country in this manner. Due to the heavy logistical and engineer requirements for the initial deploying force, only 80,000 of the 550,000 US troops were fighters. A better force mix focused on light infantry and fire support would have had a better ratio of combat soldiers, and made the CAP program a feasible approach.[268]

Eventually, the counterinsurgency effort was able to isolate the insurgent force from its base of support. However, the success of pacification from 1968 to 1971 cannot be fully understood since it coincides with the largest and most successful part of conventional operations in SVN.[269] American troops exhibited little compunction with clearing insurgents by armed force; tactical success in search-and-destroy operations convinced MACV that it was accomplishing just that. A large part of that tactical success was the heavy use of American artillery units' indirect firepower.

As with the Malayan Emergency, the requirement for tactical fires and the organization of the counterinsurgency force were the two factors that had the most influence on the employment of artillery units. As American artillery units were introduced in SVN, they were rapidly integrated into the framework and girding of search-and-destroy operations in an attempt to win the war via attrition. Given the awesome casualty-producing firepower of artillery units compared to light infantry forces at the time, this seemed like a logical role for the artillery. But if the approach to a counterinsurgency effort is a blank canvas at the outset of a campaign, there are several factors that influence the brushstrokes as the work takes form. These factors such as the terrain, constraints, and limitations on the force, and the availability of other firepower assets all influenced the employment of artillery units in SVN.

The requirement for tactical fires in SVN was about as high as can be expected in the context of a counterinsurgency effort, given that the hybrid nature of warfare had a significantly conventional quality to it. Artillery units could not be expected to reconstitute to meet the threat of a main force VC or NVA unit, then instantaneously transform to a nonstandard security or pacification role. But a nuance of this fact is that artillery units rarely had to mass fires above the battery level to meet the tactical requirements of the maneuver force.[270] As noted before, more fire missions were fired as massive amounts of unobserved H&I fires than were fired to support troops in contact. The notable exceptions were Hue and Khe Sanh, where the enemy was decimated by indirect fires whenever it massed.[271]

As a corollary, there were few chances to mass fires in the deliberate targeting of an enemy. Since the VC did not need to protect physical lines of communication or critical infrastructure, there was not a class of fixed locations to target for destruction. By definition, if a counterinsurgency force has to search for an enemy before destroying it, large preparatory fire plans do not seem logical except to secure initial movements. The seasoned VC and NVA knew this intuitively, and their notions were confirmed upon their first institutional contacts with American artillery. Senior Colonel Ha, a veteran of the Ia Drang campaign in 1965, reflected that his best method was to surprise the Americans and separate them from their firepower by maintaining close contact, or completely disengaging. He sought to win or escape before the battle became an even match.[272] Quite literally, the enemy sought to never mass long enough to allow for deliberate targeting so that American artillery units would always be supporting hasty individual calls for fire.

There were few organizational constraints on the use of artillery firepower, but many limits due to the environment in SVN. Although collateral damage was considered, it did not limit the employment of artillery units. In his introduction to the use of field artillery firepower in limited war, historian J.B.A. Bailey notes that "[i]n a limited war, the principle of 'economy of force' may still apply, but the principle of 'minimum force' is often superimposed. 'Economy of force' is desirable to husband resources, while 'minimum force' is desirable to control the effect of their product." [273] It was an especially difficult endeavor to balance these principles in Vietnam since the war was a mix of high- and low-intensity operations. Collateral damage was avoided, but accepted. MACV estimated that there were 165,000 civilian casualties and 2 million refugees as a result of mop-up operations after the Tet Offensive in Hue, Cholon, Kontum, My Tho, and Ben Tre. Although not all of these are attributable to indirect firepower, these locales had large swathes of urban sections flattened so it is a reasonable assumption that artillery was a major contributing factor.[274] There were initial restrictions to retake Hue due to the cultural significance of the citadel, but these restrictions vanished and the 1st Cavalry Division (Airmobile) Artillery alone fired 52,000 rounds into the city.[275] The normal exhortations against collateral damage in a counterinsurgency may not have been seen during these campaigns since it was relatively well known that the North Vietnamese had initiated combat in these populated areas.

The American ROE reflected this attitude on collateral damage. When the legality of unobserved H&I fires in populated areas were challenged in 1967, commanders declared them free-fire zones and issued warnings to the populations rather than expanding patrols.[276] After the Tet Offensive, MACV published an ROE in order to decrease casualties, and delineated uninhabited areas, hamlets and villages, and urban areas.[277] During *Vietnamization,* ARVN and local leaders were integrated into the process for clearing fires, which severely limited the use of artillery for H&I fires but reinforced the efforts of pacification in the rural countryside. The goal of responding to a call for fire within two minutes was seldom met in this environment. Without prejudice, *MACV Lesson Learned No. 77* regarding fire support coordination flatly stated that "the requirement for military and political clearances for artillery fire on or near populated areas has an adverse effect on the responsiveness of artillery fire."[278]

Another constraint on the employment of artillery units was the American army's own doctrine. Doctrinal field manuals made an unnecessary prioritization in roles; the artillery's primary mission in counterinsurgency doctrine was to support combat operations, and acting as security or village defense forces was a secondary mission.[279] Fire support doctrine of the time only addressed artillery support in clearing the enemy, not acting in a nonstandard role to hold and protect key terrain, infrastructure, or population centers.[280] There was a period of adjustment, since the dispersed and static positioning of general support artillery went against the theoretical advantages of massed battalion fires at decisive points.[281] Professional journals were the primary means for a discourse in doctrine, and artillery concepts did not undergo a sufficient revision until the publication of Field Manual 6-20-2 *Field Artillery Techniques* in 1970.[282]

As with Malaya, the chief physical limit on artillery fires was the terrain itself. Again, a country with thick jungle canopy, impassable marsh areas, and mountainous terrain hindered observation and mobility. The jungle was so thick in some regions that adjusting initial rounds by sound became a common but dangerous practice.[283] Beginning in 1962, artillery officer courses at Fort Sill included instruction in jungle firing operations.[284] Early techniques in orienting observers in the thick jungle included using white phosphorous to mark the fall of initial rounds, which was quickly refined to mark with smoke or ground-burst illumination.[285] Eventually, this technique would be expanded to use standard high-explosive rounds in mountainous regions to orient observers during Operation WASHINGTON GREEN in 1970.[286] The terrain also brought an advantage to American forces, since they could bring 105mm artillery within 50 meters of friendly troops in the thickest jungles.[287]

The chief limitation due to terrain was its effect on observation. As with Malaya, the restricted sight distances in the jungle contributed to quick and indecisive contacts in many regions of SVN, often before a maneuver element could leverage artillery firepower. Forward observers with the infantry were expanded to a Lieutenant, an NCO and an enlisted soldier per company in an attempt to improve tactical fire support.[288] The war also provided a testbed for field optics employed at observation posts with the ability to identify targets at a range of 10 kilometers during daylight. Advanced optics derived from naval ships provided a capability to identify targets at a range of four kilometers at night, and this new ability was used to great effect at places similar to Khe Sanh.[289]

American artillery tried to provide new sensor arrays for target location and target acquisition, but nothing was able to replace a well-trained soldier walking with the infantry. Dedicated target acquisition assets made an appearance throughout the war, to include sound-ranging sets, ground surveillance radar, and the marginally effective counter-mortar radar.[290] Large arrays of individual ground target sensors were widely used, but they were not very accurate since they were air-delivered and their final position was only an estimate. Consequently their value in leading to little more than unobserved H&I fires is debatable.[291] In 1970, the 101st Airborne Division (Airmobile) even stationed artillerymen in the cockpit with OH-6 pilots in an attempt to establish an aerial observer capability.[292] The most effective method of target acquisition was the emerging capabilities in ground surveillance radars, but even those were limited to affective use in the relatively flat delta region.[293] Still, no alternate means of observing fires replicated a trained forward observer with a good vantage point, which is why one artilleryman from the 1st Cavalry Division (Airmobile) developed a system of ropes and pulleys to get himself above the jungle canopy while on patrol.[294]

The counterinsurgency effort was task-organized in such a fashion that there was not a large demand for additional security forces. Since early efforts were focused on search-and-destroy operations, there was not an intense manpower requirement to garrison these areas. In areas that were held for pacification, this was seen as the role for ARVN forces, then the territorial forces as they stood up. If artillery units were not even forced to secure their own FSBs, it is quite a reach to think that there was an intense requirement to use them in a nonstandard security role. Given the architecture of the counterinsurgency force, there was not a need strong enough to trump the utility of indirect fires to attrit the enemy's conventional forces.

There was not a strong force pushing artillery units into those roles either because air assets were not a strong enough source of firepower to replicate artillery units in SVN. Air power during the war was efficient and plentiful, but more focused on operational and strategic effects. This efficiency was gained after years of air-ground integration in the Korean War. These techniques were consolidated in the Tactical Air Control Party, which served a battalion in the same way a forward observer served them for artillery fires. Additionally, the concept of Forward Air Controllers in OV-10 Bronco aircraft greatly enhanced the coordination and integration of air assets.[295] Air assets were also plentiful, starting with the introduction of B-57s in 1965 and followed by more high-performance jet aircraft throughout the war. From 1965 to 1968, American ground forces could rely upon roughly 300 close air-support sorties per day. But even this capacity was stressed to its breaking point during the Tet Offensive, and UH-1 gunships had to fill the gaps in support.[296]

In spite of this efficiency and availability, most aircraft were used for operational and strategic bombing. The limited B-52 strikes for tactical support had a much greater psychological effect on the VC and NVA than a direct tactical effect. The enemy also developed an early warning system to these flights, attributed to Soviet trawlers gathering intelligence around their home air bases in Guam.[297] These bombers had an incredible payload, but they were so inaccurate that they could not be used anywhere near friendly forces or populated areas.[298] Converted cargo planes such as the AC-47, AC-119, and AC-130 gave good direct-fire close air support, but they were an extremely limited asset.[299] At the time, doctrine resulting from the Howze Board put forward the notion that the Army should have responsibility for aerial fire support with their rotary-wing assets. This colored the use of fixed-wing close air support, and the efforts of the US Air Force which was institutionally dedicated to Cold War strategic bombing.[300] Attack helicopters were plentiful, but they were used primarily as escorts for airmobile operations instead of dedicated fire support platforms.[301] The two airmobile divisions' ARA battalions were used widely across SVN, but they were seen by commanders as an artillery asset to the point that they operated on the same radio nets as firing units.[302]

The relative cost of converting an artillery unit to fight in a nonstandard role in SVN was negligible, since it generally did not occur. As discussed before, artillery units did not even secure their own FSBs in many cases. One study in 1965 concluded that if properly equipped, an artillery battalion tailored for counterinsurgency operations would require an additional an additional 146 soldiers: 20 additional fire direction computers, 18 additional forward observers, and 108 infantrymen for security.[303] Training for artillerymen was focused on tasks that were new to stateside or European-based units that had been training for a Cold War conflict on the open plains of Central Europe. Mission-specific artillery skills for operations in SVN required additional training, but not re-training.[304] To train artillery officers in theater-specific intricacies, I Field Force Artillery headquarters established courses such as a six-day Fire Direction Officer Course. However, no courses existed to train officers on conventional artillery for the linear Cold War battlefield once their duties in SVN were complete.[305]

Conclusion

The American experience during the Vietnam War amply illustrates the challenge of simultaneously waging a war of conventional means and a counterinsurgency. One of the key contributing factors to this difficulty was that the enemy force coordinated both modes of warfare and retained the ability to scale violence as needed in most regions. As Bernard Fall's quote at the beginning of the chapter alludes to, the American force countered this form of warfare with their own chosen form of warfare: attrition by means of superior mobility and firepower. As such, artillery units played a leading role in this approach to a counterinsurgency effort.

Three decades after the fall of South Vietnam at the hands of a conventional invasion, American artillery units found themselves in the midst of a counterinsurgency effort again. In this case, a large conventional invasion provided the prologue instead of the epilogue to the grinding counterinsurgency. In Iraq, however, American artillery units found themselves playing a much different role with an entirely new host of challenges.

Notes

1. There is a virtually unlimited supply of historical records, accounts, critiques and analyses of Vietnam during this period. To understand the themes of this war, some of the most helpful sources are as follows. Bernard Fall's works provide an excellent context to the war: *Hell in a Very Small Place, Street Without Joy* and *The Two Vietnams*. Jeffrey Race's *War Comes to Long An* is an excellent case study in communist subversion and penetration; Mark Moyar's *Triumph Forsaken* is an incredibly detailed historical account of the earlier years in the American experience; James Willbanks' *Abandoning Vietnam* provides a great narrative for the remaining years. Sir Robert Thompson provides his insight in both *Defeating Communist Insurgency* and *No Exit From Vietnam*. General William Westmoreland's *A Soldier Reports* is an invaluable account of his decisions as the MACV Commander; Robert Komer provides similar insight for CORDS in his analysis *Bureaucracy at War*. Andrew Krepinevich gives a convincing argument why Westmoreland's approach was ill-founded in *The Army and Vietnam*; Dale Andrade's article "Westmoreland was Right: Learning the Wrong Lessons From the Vietnam War" gives an excellent counterbalance to it. The "Indochina Monographs" series produced by the CMH in the early 1980s provides lengthy analysis by senior South Vietnamese subject matter experts. The Texas Tech University archive is another excellent resource for primary source material and is available online. For artillery-specific studies, General David Ewing Ott's *Vietnam Studies: Field Artillery* is an excellent distillation of techniques from the war, and was written by then-recent practitioners at the war's conclusion at Fort Sill.

2. Some published works that directly compare Malaya and Vietnam include: *Learning to Eat Soup With a Knife: Counterinsurgency Lessons from Malaya and Vietnam* by John A. Nagl and *Conflict of Myths: The Development of American Counterinsurgency Doctrine and the Vietnam War* by Larry E. Cable. Additionally, Sir Robert Thompson frequently compares the two conflicts in both *Defeating Communist Insurgency* and *No Exit from Vietnam* which is understandable, given his prominent role and vantage point in both.

3. In Malaya, a chief advantage for the British was the fact that the CTs were almost exclusively from a different ethnic group which was easily identifiable. While this held true in the central highlands in Vietnam (ethnic minorities in the mountain tribes are different from ethnic Vietnamese), the contested regions in the overwhelming majority of South Vietnam pitted ethnic Vietnamese against ethnic Vietnamese.

4. Vietnam I Panel Discussion, US Army Command and General Staff College Art of War Scholars Seminar, 14 January 2011, Fort Leavenworth, KS.

5. The Vietnam War is not unique in this regard, as history is rife with this type of conflict: The American Revolution and the Peninsular War are but two examples.

6. Vietnam I Panel Discussion, US Army Command and General Staff College Art of War Scholars Seminar, 14 January 2011, Fort Leavenworth, KS.

7. Hoang Ngoc Lung, *Indochina Monographs: Strategy and Tactics* (Washington, DC: US Army Center for Military History, 1978), 130-131.

8. The SVN abbreviation reflects an accurate nomenclature, that of a Republic of South Viet Nam. The singular word Vietnam is an Anglicized combination of Viet (referring to the dominant ethnic group) and Nam (referring to the southern position of the nation with respect to China). Although the terms GVN (referring to the Government of Viet Nam) and SVN may be used somewhat interchangeably, in this chapter GVN will denote the political state and government, while SVN will refer to the physical setting of the conflict.

9. Ngo Quang Truong, *Indochina Monographs: Territorial Forces* (Washington, DC: US Army Center for Military History, 1981), 9.

10. Truong, *Indochina Monographs*, 10-11.

11. Bernard Fall, *The Two Vietnams*. (New York: Praeger, 1967), 6.

12. William C. Westmoreland, *A Soldier Reports* (New York: Da Capo Press, 1976), 52.

13. Fall, *The Two Vietnams*, 15.

14. Fall, *The Two Vietnams*, 54.

15. Fall, *The Two Vietnams*, 58-59.

16. Fall, *The Two Vietnams*, 67.

17. Fall, *The Two Vietnams*, 68-71; Bernard Fall, *Hell in a Very Small Place* (Philadelphia: Lippincot Press, 1966), 23.

18. Fall, *The Two Vietnams*, 78.

19. Vo Nguyen Giap, *Inside the Vietminh: Vo Nguyen Giap on Guerrilla War* (Quantico, VA: Marine Corps Association, 1962), Chapter 4. Giap's account of Dien Bien Phu is an excellent self-examination of the strengths and liabilities inherent to his style of guerrilla warfare. Although it has some tones of Marxist exhortation, it maintains a seemingly objective view towards the military aspects of the campaign.

20. George C. Herring, *America's Longest War: The United States and Vietnam, 1950-1975* (New York: McGraw-Hill, 2002), 36-39.

21. Fall, *Hell in a Very Small Place*, 482.

22. Fall, *The Two Vietnams*, 318-320.

23. Bernard Fall, *Street Without Joy* (New York: Shocken Books, 1961), 345.

24. Fall, *Street Without Joy*, 345.

25. Graham A. Cosmas, *MACV: The Joint Command in the Years of Escalation 1962-1967* (Washington, DC: US Army Center for Military History, 2005), 43-45.

26. Jeffrey Race, *War Comes to Long An: Revolutionary Conflict in a Vietnamese Province* (Berkeley, CA: University of California Press, 1972), 165-170.

27. Vo Nguyen Giap, *The Military Art of People's War*, ed. Russell Stetler (New York: Monthly Review Press, 1970), 238-239. Also refer to Lung, *Indochina*

Monographs: Strategy and Tactics, 122.

28. Fall, *The Two Vietnams,* 35-36.

29. Fall, *The Two Vietnams,* 249. Fall sees this as a manifestation of Vietnamese *personalism* rather than an immature sense of democratic institutions.

30. John H. Cushman, *Senior Officer Debriefing Report: Major General John H. Cushman, Commander, Delta Regional Assistance Command, Vietnam, 14 May 71-14 Jan 72* (Washington, DC: Department of the Army, 1972), 4.

31. Race, *War Comes to Long An,* 141.

32. Race, *War Comes to Long An,* 142-149.

33. Lung, *Indochina Monographs,* 122, 124.

34. Race, *War Comes to Long An,* 159-161.

35. Race, *War Comes to Long An,* 177-178.

36. Giap, *The Military Art of People's War,* 263. Although, like Mao, Giap's writings also doubled as exhortations to comrades in the field, the author has been unable to independently confirm if Giap is Vietnamese for *hubris*.

37. Andrew F. Krepinevich, *The Army and Vietnam* (Baltimore, MD: Johns Hopkins University Press, 1986), 191-192. This was reflected in captured VC documents during Operation Cedar Falls in January 1967.

38. Krepinevich, *The Army and Vietnam,* 168.

39. Fall, *Street Without Joy,* 347. Fall's revised edition published in 1964 contains the following passage: "The hard fact is, that , save for a few specialized anti-aircraft and anti-tank weapons and cadre personnel not exceeding perhaps 3,000 to 4,000 a year or less, the VC operation inside South Viet-Nam has become self-sustaining."

40. Westmoreland, *A Soldier Reports,* 55-57. As the MACV Commander, Westmoreland's account and postwar analysis of COSVN's role is critical to understanding the formulation of American strategy.

41. Headquarters, United States Military Assistance Command-Vietnam, *PAVN Artillery (Rocket Units)-1967* (Saigon: United States Military Assistance Command-Vietnam, 1967), 1, 34. Accessed via the Texas Tech University Vietnam Center and Archive at www.vietnam.ttu.edu/resources/collections/ (accessed 7 May 2011).

42. Lawrence H. Carruthers, "Characteristics and Capabilities of Enemy Weapons," *Artillery Trends* 45 (September 1970): 11-24.

43. Race, *War Comes to Long An,* 151.

44. Race, *War Comes to Long An,* 152.

45. Edward Lansdale, "Contradictions in Military Culture" in *The Lessons of Vietnam,* ed. W. Scott Thompson and Donaldson Frizzell (New York: Crane, Russak and Company, 1977), 42.

46. Robert Thompson, *Defeating Communist Insurgency* (London: Chatto and Windus, 1966), 121-140. Thompson devotes an entire chapter (aptly titled

"Strategic Hamlets") to the comparison and contrast of the two programs in Malaya and Vietnam. Thompson also discusses the failings of the Strategic Hamlet program in *No Exit From Vietnam*. (New York: Davis McKay Company, 1969), 169-170. Ironically, the GVN developed the first strategic hamlets on its own, several months before Thompson arrived in Vietnam. Mark Moyar, *Triumph Forsaken* (New York: Cambridge University Press, 2006), 156.

47. Robert Komer, *Bureaucracy at War: U.S. Performance in the Vietnam Conflict* (Boulder, CO: Westview Press, 1986), 138; Moyar, *Triumph Forsaken*, 158-159.

48. BA030, Vietnam Veteran, Interview by Aaron Kaufman and Dustin Mitchell, Fort Leavenworth, KS, 24 February 2011.

49. Komer, *Bureaucracy at War*, 138.

50. Fall, *Street Without Joy*, 363. While these VC battalions would have numerically outnumbered the force fielded by two ARVN divisions in the same area, it would not have the same conventional combat power in terms of coordinated firepower and logistics.

51. Peter Dale Scott, "Vietnamization and the Drama of the Pentagon Papers," in *The Pentagon Papers Volume V*, ed. Noam Chomsky and Howard Zinn (Boston: Beacon Press, 1972), 246-247.

52. Westmoreland, *A Soldier Reports*, 57.

53. Westmoreland, *A Soldier Reports*, 62-63.

54. Westmoreland, *A Soldier Reports*, 56.

55. Westmoreland served under Taylor in the 82nd Airborne Division during the Sicily campaign in World War II. Westmoreland, *A Soldier Reports*, 67-68.

56. Westmoreland, *A Soldier Reports*, 68-69. Unfortunately, Thompson does not include an account of this trip in *No Exit From Vietnam*. Westmoreland asserted that the only transferrable lesson from that counterinsurgency effort was the unity of effort embodied by the SWECs and DWECs, although no attempt was made to incorporate a similar system due to the relative weakness of district administrators within the GVN.

57. William C. Westmoreland, "A Military War of Attrition" in *The Lessons of Vietnam*, ed. W. Scott Thompson and Donaldson Frizzell (New York: Crane, Russak and Company, 1977), 65; Westmoreland, *A Soldier Reports*, 153-154 for Westmoreland's defense of pursuing a war of attrition. Westmoreland saw Vietnam in terms of a dichotomy: some military units blocked VC main force units in a conventional sense, while other units conducted the pacification effort among the coastal population centers and rural areas.

58. Truong, *Indochina Monographs*, 96.

59. Truong, *Indochina Monographs*, 26.

60. Westmoreland, *A Soldier Reports*, 57-58.

61. Mark Moyar, *Triumph Forsaken* (New York: Cambridge University Press, 2006), 310-311. North Vietnamese patrol boats attacked American naval

intelligence vessels, even after both Ho Chi Minh and Giap decided against it. While some saw this as an opportunity to address overt North Vietnamese aggression with a militarily free hand, the response was muffled by a wariness to widen the war with other Communist powers, as in Korea.

62. Moyar, *Triumph Forsaken,* 313.

63. Marshall was writing in an unofficial capacity, having retired in 1960 and assumed the life of an observer and historian-at-large in Vietnam.

64. S. L. A. Marshall, "Thoughts on Vietnam" in *The Lessons of Vietnam*, ed. W. Scott Thompson and Donaldson Frizzell (New York: Crane, Russak and Company, 1977), 65. It is unclear how Marshall viewed this decisive victory taking shape, but this probably means ground incursions into North Vietnam itself. The North Vietnamese were surely aware of the danger of being on the wrong side of another Dien Bien Phu in South Vietnam.

65. In contrast, the United States' Pacific allies contributed a significant amount of capable combat power, led by Australia, South Korea, Thailand and The Phillipines. Overall, third-country support peaked at 71,000 in 1969. Herring, *America's Longest War*, 181.

66. Westmoreland, "A Military War of Attrition," 59.

67. Lansdale, "Contradictions in Military Culture," 45.

68. BH060, Vietnam Political and Military Analyst, Interviewed by Mark Battjes, Ben Boardman, Robert Green, and Dustin Mitchell, Washington, DC, 22 March 2011; Westmoreland, *A Soldier Reports*, 87-115. Westmoreland's unique views on this subject illustrate the constant changes in leadership, and its paralyzing effect on the counterinsurgency effort. Tellingly, he titles these chapters "Chaos Unabated" and "Constant Crisis."

69. Komer, *Bureaucracy at War*, 141. Komer's view is from an important vantage point since at the time he was serving as the interim National Security Advisor to Johnson.

70. Westmoreland, "A Military War of Attrition," 59. For his reflections on this process, Westmoreland, *A Soldier Reports*, 128-129.

71. Westmoreland, "A Military War of Attrition," 62-63; Westmoreland, *A Soldier Reports*, 141-143, 145.

72. Moyar, *Triumph Forsaken*, 412-416.

73. John A. Nagl, "Counterinsurgency in Vietnam: American Organizational Culture and Learning," in *Counterinsurgency in Modern Warfare*, ed. Daniel Marston and Carter Malkasian (Oxford, UK: Osprey Publishing, 2010), 126.

74. Westmoreland, *A Soldier Reports*, 177-179.

75. Krepinevich, *The Army and Vietnam*, 173.

76. Krepinevich, *The Army and Vietnam*, 173-175.

77. Westmoreland, *A Soldier Reports*, 196-197.

78. Krepinevich, *The Army and Vietnam*, 179.

79. Vietnam I Panel Discussion, US Army Command and General Staff

College Art of War Scholars Seminar, 14 January 2011, Fort Leavenworth, KS.

80. Thompson is quoted in Dale Andrade, "Westmoreland was Right: Learning the Wrong Lessons From the Vietnam War," *Small Wars and Insurgencies* 19, no. 2 (June 2008): 163.

81. Komer, *Bureaucracy at War*, 142.

82. Komer, *Bureaucracy at War*, 142. The official title of the report was "A Program for the Pacification and Long-Term Development of South Vietnam." The first volume (PROVN I) contained the report's analysis, while the second volume (PROVN II) contained several annexes.

83. Office of The Deputy Chief of Staff for Military Operations-U.S. Department of Defense, *A Program for the Pacification and Long-Term Development of South Vietnam, vol I* (Washington, DC: Department of Defense, 1966), 1.

84. Office of The Deputy Chief of Staff for Military Operations-U.S. Department of Defense, *A Program for the Pacification and Long-Term Development of South Vietnam, vol I*, 1-2.

85. Office of The Deputy Chief of Staff for Military Operations-U.S. Department of Defense, *A Program for the Pacification and Long-Term Development of South Vietnam, vol I*, 5.

86. Office of The Deputy Chief of Staff for Military Operations-U.S. Department of Defense, *A Program for the Pacification and Long-Term Development of South Vietnam, vol II* (Washington, DC: Department of Defense, 1966), I-43.

87. Krepinveich, *The Army and Vietnam*, 182. In a pre-emptive warning against this notion of making the regime look like a puppet government, PROVN II stated that "[t]he Vietnamese can be talked to frankly and honestly about the situation in their country. Creating a facade of success destroys the credibility of the US support effort by casting Americans as either insincere or ignorant of local conditions." Although PROVN illustrated why operations were failing to achieve a strategic effect in SVN, the army suppressed the report lest any resources be reallocated toward the support of pacification or civic action.

88. Komer, *Bureaucracy at War*, 144.

89. Krepinevich, *The Army and Vietnam*, 185-186.

90. Giap, *The Military Art of People's War*, 253.

91. Richard Hunt, *Pacification: The American Struggle for Vietnam's Hearts and Minds* (Colorado: Westview Press, 1995) 82.

92. Hunt, *Pacification*, 83-84.

93. Westmoreland, *A Soldier Reports*, 212.

94. Hunt, *Pacification*, 83.

95. Westmoreland, *A Soldier Reports*, 205-206.

96. Krepinevich, *The Army and Vietnam*, 187.

97. Krepinevich, *The Army and Vietnam*, 188.

98. Krepinevich, *The Army and Vietnam*, 190.

99. Krepinevich, *The Army and Vietnam*, 187; Westmoreland, *A Soldier Reports*, 214.

100. Thompson, *Defeating Communist Insurgency*, 102.

101. Hunt, *Pacification*, 86-87.

102. Hunt, *Pacification*, 88, 91.

103. Hunt, *Pacification*, 90.

104. Hunt, *Pacification*, 90. This included ratings and evaluations for the subordinates in both cases.

105. Hunt, *Pacification*, 89.

106. BA030, Vietnam Veteran, Interview.

107. Vietnam II Panel Discussion, US Army Command and General Staff College Art of War Scholars Seminar, 18 January 2011, Fort Leavenworth, KS.

108. Westmoreland, *A Soldier Reports*, 230-231.

109. Jesse Faugstad, "No Simple Solution," *Military Review* (July-August 2010): 34-35.

110. Headquarters United States Military Assistance Command-Vietnam, *RF-PF Handbook for Advisors* (Saigon: United States Military Assistance Command-Vietnam, 1969), 6.

111. Faugstad, "No Simple Solution," 39, 41.

112. Headquarters, United States Military Assistance Command-Vietnam, *Command History 1968 Volume I* (Saigon: United States Military Assistance Command-Vietnam, 1969), 376.

113. Krepinevich, *The Army and Vietnam*, 239-240.

114. Marshall, "Thoughts on Vietnam," 53.

115. Westmoreland, *A Soldier Reports*, 310-311.

116. Westmoreland, *A Soldier Reports*, 331.

117. Westmoreland, *A Soldier Reports*, 328-330.

118. Robert H. Scales, *Firepower in Limited War* (Novato, CA: Presidio Press, 1995), 120.

119. Westmoreland, *A Soldier Reports*, 338-339.

120. Krepinevich, *The Army and Vietnam*, 249-250.

121. Westmoreland, *A Soldier Reports*, 334.

122. Truong, *Indochina Monographs*, 96; Westmoreland, *A Soldier Reports*, 332.

123. Truong, *Indochina Monographs*, 96.

124. Westmoreland, "A Military War of Attrition," 69.

125. Krepinevich, *The Army and Vietnam*, 245.

126. Krepinevich, *The Army and Vietnam*, 247.

127. Eric Bergerud, *The Dynamics of Defeat: The Vietnam War in Hau Nghia Province* (Oxford, UK: Westview Press, 1991), 223.

128. Tran Dinh Tho, *Indochina Monographs: Pacification* (Washington, DC: US Army Center for Military History, 1978), 183.

129. Bergerud, *The Dynamics of Defeat*, 224-226.

130. Truong, *Indochina Monographs*, 94.

131. Bergerud, *The Dynamics of Defeat*, 237.

132. Bergerud, *The Dynamics of Defeat*, 226.

133. Bergerud, *The Dynamics of Defeat*, 234.

134. Krepinevich, *The Army and Vietnam*, 251.

135. James Willbanks, *Abandoning Vietnam* (Lawrence, KS: University of Kansas Press, 2004), 21.

136. Westmoreland, *A Soldier Reports*, 252; Bergerud, *The Dynamics of Defeat*, 241.

137. Bergerud, *The Dynamics of Defeat*, 50.

138. Hunt, *Pacification*, 269-271. Clifford approved initiatives in an attempt to ensure that the re-emerging advisory role was given equal weight as the departing commanders' role in military personnel systems. Policies including General Officer letters of recommendation, family housing for stateside dependents, advanced civil schooling after tours, and guaranteed consideration for early command made the assignment an attractive assignment, especially for career officers who did not gain combat experience from 1965 to 1968. However, these incentive policies could never compete with the institutional emphasis on command experience.

139. Bergerud, *The Dynamics of Defeat*, 244-245. The "tethered goat" tactics involves portraying a weak and defenseless unit (in this scenario, often a PF platoon) to bait an enemy attack, then destroying the enemy with a previously hidden, superior element (a combined American and ARVN force with generous amounts of firepower at their disposal).

140. Bergerud, *The Dynamics of Defeat*, 249.

141. Bergerud, *The Dynamics of Defeat*, 246, 252.

142. Westmoreland, *A Soldier Reports*, 296. This was corroborated specifically in BH060, Interview.

143. Lewis Sorley, *Vietnam Chronicles: The Abrams Tapes 1968-1972* (Lubbock, TX: Texas Tech University Press, 2004), 318-320. A sampling of transcripts from Abrams' meetings in December 1969 illustrates both the improvement in ARVN and the drastic decrease in main force infiltration into SVN. Sorley served as Abrams' Chief of Staff during the war, so this record may be prone to a biased presentation of the material.

144. Truong, *Indochina Monographs*, 40.

145. Hunt, *Pacification*, 273.

146. Graham A. Cosmas, *MACV: The Joint Command in the Years of Withdrawal 1968-1973* (Washington, DC: US Army Center for Military History, 2005), 320-324.

147. Cushman, *Senior Officer Debriefing Report*, 9.

148. Cosmas, *MACV*, 345, 351.

149. Cosmas, *MACV,* 385.

150. Herring, *America's Longest War*, 324-325.

151. Herring, *America's Longest War*, 326.

152. Hunt, *Pacification*, 274.

153. Vietnam II Panel Discussion, US Army Command and General Staff College Art of War Scholars Seminar, 18 January 2011, Fort Leavenworth, KS. A veteran of the CORDS program from the delta region assessed that the need to mentor RFs and PFs had passed as early as 1972 in his sector.

154. Vo Nguyen Giap and Van Tien Dung, *How We Won The War* (Philadelphia, PA: RECON Publications, 1976), 41. Giap's views were corroborated during BH060, Interview, by an American political-military expert, who stated that "if the North Vietnamese had never come in, the south would never have fallen because the regional forces and the military could take care of the Vietcong. We had trained them to that point."

155. Herring, *America's Longest War*, 332.

156 . Willbanks, *Abandoning Vietnam*, 281, 284.

157. Westmoreland, *A Soldier Reports*, 18-21, 40, 70.

158. Krepinevich, *The Army and Vietnam*, 164.

159. Boyd L. Dastrup, *King of Battle: A Branch History of the U.S. Army's Field Artillery* (Washington, DC: Center of Military History, 1992), 284. Additionally, the preface to Scales' *Firepower in Limited War* provides an excellent distillation of this concept.

160. Westmoreland, *A Soldier Reports*, 282.

161. Janice E. McKenney, *The Organizational History of Field Artillery 1775-2003* (Washington, DC: US Army Center for Military History, 2007), 269; John J. McGrath, *Fire for Effect: Field Artillery and Close Air Support in the US Army* (Fort Leavenworth, KS: Combat Studies Institute Press, 2007), 117.

162. Scales, *Firepower in Limited War*, 61.

163. Bruce I. Gudmundsson, *On Artillery* (Westport, CT: Praeger, 1993), 151.

164. E. D. Smith, *Counter-Insurgency Operations 1: Malaya and Borneo* (London: Ian Allan, 1985), 68.

165. Townsend A. Van Fleet, "105 in the Jungle," *Artillery Trends* 26 (October 1962): 19-26.

166. Dastrup, *King of Battle*, 277.

167. Gudmundsson, *On Artillery*, 151.

168. Patrick J. Lindsay, "Counterguerrilla Warfare," *Artillery Trends* 29 (February 1964): 13.

169. Dastrup, *King of Battle*, 280-281. These were not a purely American invention, as the British RA has already pioneered single-gun raids in Borneo. Refer to Smith, 68-69.

170. Gudmundsson, *On Artillery*, 152; Scales, 91.

171. Ian Horwood, *Interservice Rivalry and Airpower in the Vietnam War* (Fort Leavenworth, KS: Combat Studies Institute Press, 2006), 129; Dastrup, 283.

172. US Army Artillery and Missile School, "Aerial Rocket Artillery," *Artillery Trends* 37 (January 1967): 42.

173. David Ewing Ott, *Vietnam Studies: Field Artillery, 1954-1973* (Washington, DC: Department of the Army, 1975), 75-80.

174. US Army Artillery and Missile School, "Riverine Artillery," *Artillery Trends* 39 (January 1968): 14-24.

175. McKenney, *The Organizational History of Field Artillery*, 270; Ott, 172.

176. McGrath, *Fire for Effect*, 116; Ott, *Vietnam Studies*, 172.

177. The Beehive round (designated the XM546) was specifically developed for anti-personnel defense in the direct fire mode. As a reflection of the intense requirements for local defense, artillery units had more Beehive rounds in their Unit Basic Load than any other munition except the standard High Explosive shell. Headquarters 25th Infantry Division Artillery, *Operational Report of 25th Inf Div Arty for Period Ending 31 Jul 68* (Washington, DC: Department of the Army, 1968), 9. Accessed via the Texas Tech University Vietnam Center and Archive at www.vietnam.ttu.edu/resources/ collections/ (accessed 7 May 2011).

178. The Killer Junior technique was developed in-country and used mechanical time fuzes set to burst 30 feet high, at a range less than a kilometer. It was particularly useful once the enemy determined that they could avoid much of the Beehive fragmentation by laying prone or behind minimal cover. The Killer Junior technique was used for 105mm and 155mm artillery, and a Killer Senior technique was eventually developed for the 8-inch howitzer. Ott, *Vietnam Studies*, 61.

179. Modern artillery uses the mil as its graduated unit of measure for firing data. There are 6,400 mils in a complete circle.

180. Nathaniel W. Foster, "Speed Shifting the 155mm Howitzer, Towed: The Evolution of an Idea," *Artillery Trends* 37 (January 1967): 17-22.

181. US Army Artillery and Missile School, "Lessons Learned in Vietnam: 6400-mil Traverse for the 175mm Gun and 8-Inch Howitzer," *Artillery Trends* 37 (January 1967): 71.

182. US Army Artillery and Missile School , "Simple Solutions for 6400-mil Charts," *Artillery Trends* 37 (January 1967): 11-16.

183. Ott, *Vietnam Studies*, 42-43.

184. Headquarters 25th Infantry Division Artillery, *Operational Report of 25th Inf Div Arty for Period Ending 31 Jul 68*, 1-2. Accessed via the Texas Tech University Vietnam Center and Archive at www.vietnam.ttu.edu/resources/collections/ (accessed 7 May 2011). This after-action review shows the reorganization and switches in command support relationships during the response to increased attacks on Saigon in 1968. For a comparison of the responsibilities inherent to divisional and area support artillery headquarters, see Ott, *Vietnam Studies*, 45-47.

185. McKenney, *The Organizational History of Field Artillery*, 268-269.

186. Department of the Army, Field Manual 3-16, *Counterguerrilla Operations* (Washington, DC: Department of the Army, 1963), 89.

187. J. B. A. Bailey, *Field Artillery and Firepower* (Oxford, UK: Oxford Military Press, 1989), 247.

188. Dastrup, *King of Battle*, 285.

189. Headquarters 173rd Airborne Brigade, *Combat After Action Report-Operation Niagara/Cedar Falls, Period 5-26 Jan 1967* (Washington, DC: Department of the Army, 1968), 51.

190. Headquarters 173rd Airborne Brigade, *Combat After Action Report-Operation Niagara/Cedar Falls, Period 5-26 Jan 1967*, 54.

191. Headquarters 173rd Airborne Brigade, *Combat After Action Report-Operation Niagara/Cedar Falls, Period 5-26 Jan 1967*, 57.

192. Dastrup, *King of Battle*, 285.

193. Headquarters 173rd Airborne Brigade, *Combat After Action Report-Operation Junction City I* (Washington, DC: Department of the Army, 1967), 5; Ott, *Vietnam Studies*, 116.

194. Headquarters 173rd Airborne Brigade, *Combat After Action Report-Operation Junction City I*, 5, 34; Ott, *Vietnam Studies*, 116.

195. Ott, *Vietnam Studies*, 117.

196. Headquarters 1st Infantry Division, *After Action Report-Operation Junction City* (Washington, DC: Department of the Army, 1967), 80.

197. Headquarters 1st Infantry Division, *After Action Report-Operation Junction City*, 81.

198. Headquarters 173rd Airborne Brigade, *Combat After Action Report-Operation Junction City I*, 39.

199. Ott, *Vietnam Studies*, 116.

200. Headquarters 173rd Airborne Brigade, *Combat After Action Report-Operation Junction City I*, 39. When firing artillery for extended operations, the long-term management of ammunition and propellant lots is a critical task. Since these are made and handled differently, they will impart different variations on the piece's muzzle velocity, and cause a variation in range. By resourcing units with a limited number of different lots, they can manage these variations with ease and ensure accurate predicted fires.

201. Bailey, *Field Artillery and Firepower*, 245.

202. Westmoreland, *A Soldier Reports*, 337-338, 340.

203. Bailey, *Field Artillery and Firepower*, 243-245. This patterned "firepower box" technique is a common theme throughout analyses on modern artillery to which many historians pay homage. Diagrams showing the technique appear in Bailey's *Field Artillery and Firepower*, 242, Scales' *Firepower in Limited War*, 130, and Ott's *Vietnam Studies: Field Artillery*, 152.

204. Ott, *Vietnam Studies*, 157.

205. Kevin M. Boylan, "The Red Queen's Race: Operation Washington Green and Pacification in Binh Dinh Province, 1969-70," *The Journal of Military History* 73 (October 2009): 1196.

206. Headquarters 4th Infantry Division Artillery, *Operational Report-Lessons Learned, Headquarters 4th Infantry Division Artillery, Period Ending 31 Jan 1970* (Washington, DC: Department of the Army, 1970), 2-3, 10.

207. Headquarters 4th Infantry Division Artillery, *Operational Report-Lessons Learned, Headquarters 4th Infantry Division Artillery, Period Ending 31 Jan 1970* 19.

208. Boylan, "The Red Queen's Race," 18, 32.

209. Headquarters, 101st Airborne Division (Airmobile), *Combat Operations After Action Report, Operation Randolph Glen* (Washington, DC: Department of the Army, 1970), 6.

210. Headquarters, 101st Airborne Division (Airmobile), *Combat Operations After Action Report, Operation Randolph Glen*, 23. For examples, see pages 11 through 12, 14, and 16 for accounts of individual unit displacements to FSBs.

211. Headquarters, 101st Airborne Division (Airmobile), *Combat Operations After Action Report, Operation Randolph Glen*, 23, 26. The author suspects that classifying most fire missions as "suspected targets" was a clever way to use unobserved H&I fires when a unit was not supposed to use unobserved H&I fires.

212. Gudmundsson, *On Artillery*, 152.

213. Dastrup, *King of Battle*, 278; Ott, 22-24.

214. Ott, *Vietnam Studies*, 26.

215. Ott, *Vietnam Studies*, 29-30.

216. McKenney, *The Organizational History of Field Artillery*, 283.

217. Headquarters US Army Section Military Assistance Advisory Group, *Lessons Learned Number 31: Artillery Organization and Employment in Counter Insurgency* (Washington, DC: Department of the Army, 1963), 2-3, 5-6.

218. Ott, *Vietnam Studies*, 191.

219. Ott, *Vietnam Studies*, 199-200. General Ott's monograph on the *Vietnamization* of ARVN artillery that forms part of this book was reprinted as a stand-alone article in *Field Artillery* magazine, based on the emerging effort to adapt Afghanistan's Soviet-era artillery forces. David E. Ott, "Vietnamization: FA

Assistance Programs," *Field Artillery* (January-February 2007): 34-41.

220. Bailey, *Field Artillery and Firepower*, 246.

221. Headquarters, 101st Airborne Division (Airmobile), *Operational Report-Lessons Learned, 101st Airborne Division (Airmobile), Period Ending 31 October 1970* (Washington, DC: Department of the Army, 1970), 49.

222. Ott, *Vietnam Studies*, 212-214.

223. Bailey, *Field Artillery and Firepower*, 247.

224. Cushman, *Senior Officer Debriefing Report*, 28.

225. Westmoreland, *A Soldier Reports*, 186. Due to logistical shortages in 1966, Westmoreland briefly limited ARVN guns to two rounds a day without restricting US guns. To be fair, US forces were engaged in heavier fighting at the time.

226. Lindsay, "Counterguerrilla Warfare," 15.

227. Headquarters Military Assistance Command-Vietnam, *RF-PF Handbook for Advisors*, 29, 31.

228. Department of the Army, Field Manual 31-73, *Advisor Handbook for Counterinsurgency* (Washington, DC: Department of the Army, 1965), 85-86.

229. Truong, *Indochina Monographs*, 99.

230. Headquarters, Military Assistance Command-Vietnam, *RF-PF Handbook for* Advisors, 32.

231. Truong, *Indochina Monographs*, 98.

232. Ott, *Vietnam Studies*, 216.

233. Truong, *Indochina Monographs*, 43.

234. Vietnam II Panel Discussion, US Army Command and General Staff College Art of War Scholars Seminar, 18 January 2011, Fort Leavenworth, KS.

235. Westmoreland, "A Military War of Attrition," 64.

236. Andrade, "Westmoreland was Right," 150.

237. Westmoreland, *A Soldier Reports*, 69 and 82.

238. Tho, *Indochina Monographs*, 184.

239. Krepinevich, *The Army and Vietnam*, 168.

240. Westmoreland, *A Soldier Reports*, 150-151.

241. Cosmas, *MACV*, 225.

242. Thompson, *Defeating Communist Insurgency*, 165.

243. Andrade, "Westmoreland was Right," 147.

244. Sorley, *Vietnam Chronicles*, 376.

245. Andrade, "Westmoreland was Right," 174.

246. Truong, *Indochina Monographs*, 137.

247. Komer, *Bureaucracy at War*, 147.

248. Andrade, "Westmoreland was Right," 146.

249. William Colby, *Honorable Men: My Life in the CIA* (New York: Simon and Schuster, 1978), 286.

250. Bergerud, *The Dynamics of Defeat*, 242.

251. Allowing, of course, for an invariable statistical incidence of Harvard-educated lawyers who were raised on a farm in Iowa.

252. Don Luce, "Tell Your Friends that We're People," in *The Pentagon Papers Volume V*, ed. Noam Chomsky and Howard Zinn (Boston: Beacon Press, 1972), 91.

253. Bergerud, *The Dynamics of Defeat*, 225; Truong, 113.

254. Bergerud, *The Dynamics of Defeat*, 223-224.

255. Willbanks, *Abandoning Vietnam*, 41-42.

256. Willbanks, *Abandoning Vietnam*, 279-280.

257. Komer, *Bureaucracy at War*, 149-150.

258. Hunt, *Pacification*, 83, 90.

259. Hunt, *Pacification*, 278.

260. Vietnam I Panel Discussion, US Army Command and General Staff College Art of War Scholars Seminar, 14 January 2011, Fort Leavenworth, KS.

261. Hunt, *Pacification*, 277, 279.

262. Krepinevich, *The Army and Vietnam*, 168.

263. Fall, *Street Without Joy*, 347.

264. Westmoreland, *A Soldier Reports*, 182.

265. Krepinevich, *The Army and Vietnam*, 170-171.

266. Thompson, *Defeating Communist Insurgency*, 169-171.

267. BA030, Vietnam Veteran, Interview. This veteran stated that the Strategic Hamlet program destroyed his intelligence network and brought many "strangers" into the area, allowing the VC to insert their sleeper agents in the hamlet.

268. Krepinevich, *The Army and Vietnam*, 176.

269. Komer, *Bureaucracy at War*, 152.

270. Gudmundsson, *On Artillery*, 151.

271. Westmoreland, *A Soldier Reports*, 340.

272. Nagl, "Counterinsurgency in Vietnam: American Organizational Culture and Learning," 126; Scales, *Firepower in Limited War*, 73.

273. Bailey, *Field Artillery and Firepower*, 235.

274. Bailey, *Field Artillery and Firepower*, 249.

275. Ott, *Vietnam Studies*, 142-143.

276. Bailey, *Field Artillery and Firepower*, 248.

277. Ott, *Vietnam Studies*, 173-175.

278. Ott, *Vietnam Studies*, 18.

279. Department of the Army, Field Manual 31-73: *Advisor Handbook for Counterinsurgency*, 84.

280. Department of the Army, Field Manual 3-16: *Counterguerrilla Operations*, 88-92.

281. McKenney, 267.

282. Andrew J. Birtle, *U.S. Army Counterinsurgency and Contingency Operations Doctrine 1942-1976* (Washington, DC: Center of Military History, 2006), 447.

283. Charles W. Jackson, "Adjust By Sound," *Artillery Trends* 39 (January 1968): 30-34.

284. Ott, *Vietnam Studies*, 23. This must have been a challenge for the Fort Sill instructors to replicate the jungle environment since Southwestern Oklahoma lacks moisture, foliage, and any kind of night life.

285. Dennis I. Walter, "First Round Smoke," *Artillery Trends* 39 (January 1968): 25-27; Ott, *Vietnam Studies*, 85.

286. Headquarters 4th Infantry Division Artillery, *Operational Report-Lessons Learned, Headquarters 4th Infantry Division Artillery, Period Ending 31 Jan 1970*, 6.

287. Scales, *Firepower in Limited War*, 88.

288. McKenney, *The Organizational History of Field Artillery*, 281.

289. Gudmundsson, *On Artillery*, 152.

290. McKenney, *The Organizational History of Field Artillery*, 282.

291. Scales, *Firepower in Limited War*, 119-120.

292. Headquarters, 101st Airborne Division (Airmobile), *Operational Report-Lessons Learned, 101st Airborne Division (Airmobile), Period Ending 31 October 1970*, 49.

293. Ira A. Hunt, *The 9th Infantry Division in Vietnam: Unparalleled and Unequaled* (Lexington, KY: University of Kentucky Press, 2010), 64-65. Major General Hunt (formerly Division Chief of Staff for the 9th Infantry Division) describes the A/N-TPS 25 and the A/N-PPS 5 radars which were leveraged to vector maneuver units towards points of origin for insurgent artillery fires on Saigon.

294. Ott, *Vietnam Studies*, 130.

295. McGrath, *Fire for Effect*, 117-118; Horwood, *Interservice Rivalry and Airpower in the Vietnam War*, 125, 127.

296. Horwood, *Interservice Rivalry and Airpower in the Vietnam War*, 126, 130.

297. For a first-hand account of this effective effort, refer to Truong Nhu Tang, *A Vietcong Memoir* (San Diego, CA: Harcourt-Brace Jovanovich, 1985),

168-170. See also: Lung, 107-108.

298. Bailey, *Field Artillery and Firepower*, 246.

299. McGrath, *Fire for Effect*, 120.

300. Horwood, *Interservice Rivalry and Airpower in the Vietnam War*, 119.

301. McGrath, *Fire for Effect*, 121-122; Horwood, 122-123.

302. Scales, *Firepower in Limited War*, 106.

303. Kenneth S. Heitzke, "Field Artillery Mission, Weapon, and Organization for Counterguerilla Warfare" (Master's Thesis, US Army Command and General Staff College, Fort Leavenworth, KS, 1965), 109.

304. Lindsay, "Counterguerrilla Warfare," 12.

305. Ott, *Vietnam Studies*, 133.

Chapter 5
Operation Iraqi Freedom

> Do not try to do too much with your own hands. Better the Arabs do it tolerably than that you do it perfectly. It is their war, and you are to help them, not to win it for them. Actually, also, under the very odd conditions of Arabia, your practical work will not be as good as, perhaps, you think it is.
>
> — T. E. Lawrence

Similar to the American experience in Vietnam, the counterinsurgency effort during Operation IRAQI FREEDOM (OIF) was not a diametrically opposed insurgency-counterinsurgency dynamic; it was also an amalgam of several forms of warfare. Insurgency, communal conflict, and terrorism all emerged in Iraq.[1] The history of OIF reinforces the notion that while war among the people has a chiefly political element, it is still warfare.[2] Insurgency took the dual forms of a rebellion of disaffected groups after the collapse of an authoritarian government, and a rebellion against an intervening authority. Communal conflict took the form of sectarian and ethnic warfare. Terrorism took the form of activities that sought to further transnational extremist goals.[3] One of the greatest challenges facing the American-led counterinsurgency effort in Iraq was that attacking one of these problem sets indirectly affected the other two problem sets.

Through an adaptation of tactics and an evolution of strategy that pragmatically secured the Iraqi population and defeated the drivers of instability, it is fair to assert "that we and the Iraqis together pulled their society back from the brink of total collapse."[4] Although OIF availed reversible gains in security and political harmony which will take years to assess, this notion represents the eventual effectiveness of an intense counterinsurgency in Iraq. Artillery units played an integral part in reinforcing the early ill-suited approaches, and directly supported the eventual success of pulling the population back from the precipice of societal collapse, in order to address security and communal conflict with enduring outcomes. The steady evolution of employing artillery units to provide indirect fires, additional forces for security missions, or a combination of the two nearly parallels the arc of counterinsurgent capability in Iraq in a larger sense. The organizational conversion cost of switching these mission profiles was not only the greatest factor in how they were employed, but also the greatest long-term effect on the units themselves.

Background

Modern Iraq covers 437,072 square kilometers, an area slightly larger than the state of California. It shares relatively accessible borders with Iran, Kuwait, Saudi Arabia, Jordan, Syria, and Turkey and has a seaport on the Persian Gulf in the extreme southeast of the country. In terms of physical geography, the land is generally a vast desert landscape periodically interrupted by fertile river valleys and rocky escarpments. The northern half of the country does have varied elevations, with hill country and mountains dominating the extreme northern areas.[5] Unlike Malaya and Vietnam, the terrain would not play a large role in the counterinsurgency when compared to the critical demographics of the country.

Iraq's population of 24.6 million had several distinctions along sectarian, ethnic, and urban-rural lines. Along religious lines, the population was roughly 60 percent Shia, 35 percent Sunni, with traces of Christian and other religious communities. Ethnically, the Arab population accounted for nearly 80 percent and Kurds an additional 15 percent, and a few socially isolated minority communities of Turkmen, Assyrians, and other groups.[6] The division along sectarian and ethnic lines colored initial American perceptions of the population in Iraq and drove these divisions further. This may have been driven by the disaffected Shia exiles and nationalistic Kurds that influenced early US plans for civil redevelopment in Iraq. Consequently, most initial approaches to counterinsurgency were couched in the terms of Shia and Sunni or Arab and Kurd.[7]

An equally important demographic aspect of Iraqi society was the conflict between urban and rural populations. Iraq had many large metropolitan centers, to include the capital city of Baghdad that had more than 5.6 million residents. Mosul in the north and Basra in the south were the other two major metropolitan areas, both cities with well over 1 million inhabitants and relatively modern infrastructure. Kurdish population centers in the north included Irbil (839,600), Kirkuk (728,800), and As Sulaymaniyah (643,200); Shia Arab cities in the south were An Najaf (563,000), Karbala (549,700), and An Nasiriyah (535,100). Cities in the Sunni Arab heartland were considerably smaller in size, with Fallujah and Ramadi along the Euphrates River Valley, and Balad, Samarra, Tikrit, and Bayji along the Tigris River Valley.[8] These urban centers were linked by interconnecting highway networks, and the desert hinterland in between the cities held little value to anyone aside from the irrigated river valleys. The traditional tribal structure held sway over most political, interpersonal, and familial relationships within Iraq, but the effect of the tribal system was greatly diluted in the larger urban areas. To a certain degree, urban Sunni and Shia nationalists had more in common than urban and rural Islamists from the same sect or ethnicity.[9]

After the Ottoman Empire's defeat in World War I, the British occupation lent itself to a mandate from the League of Nations.[10] Iraq gained formal independence in 1932 but many British nationals stayed on in key government positions, and the military maintained RAF air bases and the training of Iraqi army units.[11] Through 1968 the country experienced an endless cycle of re-formed parliamentarian governments, internal political machination, and outright coups. The Ba'ath Party was founded in Syria in the 1940s and established an office in Iraq in 1951, initially appealing to the educated and working classes of both sects.[12] In 1964, the regime jailed Saddam Hussein after an abortive coup attempt by the Ba'athists; however, he emerged to lead an even stronger Ba'ath Party in 1966. When the dust settled after a successful coup in 1968, the Ba'ath Party held control in a confederation and the ensuing regime established itself as a nominally Ba'athist government.[13]

The Iran-Iraq War began in September 1980 when Iraq declared the entire Shatt al-Arab region to be rightfully theirs and launched pre-emptive strikes to capture towns in the region.[14] Throughout the war, which both governments cast as an existential struggle for survival, both sides invoked historical themes of Arab and Persian conflict from battles dating as far back as the seventh century. Hussein used the conflict as a means to further weaken the Shia populations within Iraq, making a clear communal division.[15] The Iran-Iraq War stressed Iraq's economy to the breaking point, and debt repayment constituted 50 percent of Iraq's oil income by 1990.[16] Hussein stretched himself strategically and invaded Kuwait for the immediate economic gain and to intimidate other weakly-armed Persian Gulf Arab states for future concession. In the face of a huge international coalition formed to oust Iraqi ground forces from Kuwait with American firepower and backing from the UN, Hussein changed his focus from "advantageous withdrawal" to "survivable withdrawal" in what would come to be called the Gulf War.[17]

To Hussein's credit, he achieved this "survivable withdrawal" as three critical American miscalculations enabled this strategic escape and domestic re-consolidation. First, American forces encouraged Shia and Kurdish minorities to stand up to the regime but did not directly support them. Additionally, political analysts assumed that Hussein's fall from power was inevitable. Finally, the coalition military failed to destroy the core of his power (the 80,000-strong Republican Guard).[18] In a parallel to future American post-combat efforts, one US Army officer who served in the Gulf War reflected that "everybody thought that the thing was over. I find that as one expression of this tendency to think that good operations fix the problem and that therefore there's no need to think beyond when the shooting stops."[19] The coalition established no-fly zones within the northern and southern regions in Iraq in a failed effort to protect the Kurdish and Shia Arabs respectively. Diplomatic efforts focused on sanctions and efforts to isolate the Hussein regime. This piecemeal strategy reflected the conventional wisdom at the time that a contained Iraq was better than a chaotic Iraq. Hussein walked a fine line but never directly opposed the American military through the next decade.[20]

Almost immediately after the Al Qaeda's terrorist attacks against the United States in 2001, the US Deputy Secretary of Defense Paul Wolfowitz made a case for a pre-emptive attack on Iraq.[21] The Bush administration used a doctrine of pre-emption to justify a military attack on Iraq due to intelligence analysis casting the Hussein regime as an intersection of Weapons of Mass Destruction (WMDs) and state-supported terrorism.[22] The administration based their doctrine of pre-emption on Cold War theories that used an imminent and existential communist bloc as a threat model. Iraq presented an imminent yet marginally existential threat, while Al Qaeda presented a marginally imminent yet existential threat to America.[23] This would have a lasting effect on the ensuing counterinsurgency effort in Iraq since the American military entered the war with limited public support for regime change and civil reconstruction directly related to the effects of war.

After gaining a nominal international backing from the UN and forming a coalition, the American military led a multidivisional ground offensive in March and April of 2003.[24] This campaign sought to force the early collapse of the Ba'athist regime, but on the third day plans to vertically envelop the regime with Special Operations, Rangers, and Airborne forces were abandoned when mechanized ground forces made incredible gains on the drive from Kuwait to Baghdad.[25] Although coalition forces took many measures not to be seen as occupiers, they were the only form of security in most areas immediately after the ground war concluded.[26]

The Insurgency

Within this relative void of locally recognized security, several groups began to emerge with deeply held grievances against the provisional authority and Iraqi central governance in general. CIA estimates warned of the propensity for Arab Muslims to view a foreign army as occupiers in January 2003. This lent credence to the effectiveness of Al Qaeda's declaration of *jihad* in 1998, which cast *jihad* as a personal duty against nonbelievers in invaded countries.[27] Although this declaration only affected an extreme minority of violent extremist Muslims in Iraq, the opening phase of terrorism and insurgency only requires this small minority. That same small minority of extremists held the worldview that the *mujahedin* defeat of the Soviet Red Army in Afghanistan directly caused the collapse of the antireligious international Communist structure, and sought to achieve the same effect against America in Iraq.[28] By November 2003, the Iraqi public began to compare the coalition to Israeli military forces during the second *intifada*.[29]

Grievances that were specific to the Sunni and Shia communities also emerged early in the insurgency. Sunni Arabs felt marginalized almost immediately after Hussein's predominately Sunni regime fell, chiefly personified by a Shia-dominated Iraqi Governing Council in the provisional government. The old Iraqi Army and Ba'ath Party apparatus were the two main bastions of Sunni societal, political, and economic power, and their forced dissolutions exacerbated the growing sense of marginalization in the new Iraq.³⁰ The Shia communities' grievances were much more localized. In 2003, British forces identified that these communities faced an unaddressed lack of essential services, and local Shia leaders openly questioned whether supporting the provisional government was the right way ahead.³¹ Economic necessity and resource scarcity drove some violence in the early stages of the insurgency, but by 2007 interviews with detainees indicated that insurgents were using pay as additional income for comfort and luxury items, not necessities.³²

Several insurgent groups rose from these grievances within Iraq. In contrast to Malaya and Vietnam, there was no unifying political structure or even aligned objectives. As early as June 2003, commanders on the ground saw the insurgency as a patchwork of many different elements. Colonel Peter R. Mansoor, the commander of 1st Brigade Combat Team (BCT), 1st Armored Division in Baghdad realized that the unit was "fighting former regime-backed paramilitary groups, Iranian-based opposition, organized criminals and street thugs" simultaneously.³³ By this time, at least nine distinct semi-organized groups, both Sunni and Shia, emerged across Iraq.³⁴

Within the Sunni base of support, insurgent groups relied on tacit support from tribal sheikhs and politicians, leaving no moderate element to which the counterinsurgency effort could appeal.³⁵ These regional or local groups alternatively competed and cooperated with the power structure of the former regime, who saw the aftermath of the ground war as just another chapter in the long struggle for Ba'ath Party supremacy in the region.³⁶ Initially, the former regime element had a relative strength; since the country was not cleared in a military sense it allowed the Special Republican Guard and the Saddam Fedayeen paramilitary to dissolve into Sunni and mixed-sect population centers, fighting coalition forces along the way.³⁷ Early attempts to kill senior regime leaders struck precisely, but not when they were physically present and most leaders escaped Baghdad before coalition forces reached the city.³⁸ Although politically connected among Sunni elites, this element would quickly atrophy due to its artificial means of support prior to the invasion. A CIA report in 2003 revealed that Hussein himself did not have a plan for a protracted guerrilla campaign; his sons Uday and Qusay were on the run with "meager support" instead of actively coordinating operations.³⁹

Within the Shia base of support, the Badr Corps group established a support network based on their previous experience in irregular warfare against the Ba'ath Party. Shia populism in Baghdad and the southern Shia strongholds led to the consolidation of forces loyal to Moqtada al-Sadr in August 2003, the Jaysh al-Mehdi (JAM).⁴⁰ In addition to the sect-specific insurgent groups, both Sunni and Shia tribes exerted their power through local security and protection for some extra-governmental activities that the coalition viewed as criminal activity.⁴¹

In addition to locally and regionally based insurgent groups, international terrorist groups saw a strategic opportunity in Iraq. Foreign fighters made one of their first appearances during an engagement with Special Operations forces during the ground war, an early sign that there was a small but willing pool of volunteers to infiltrate and fight Westerners in Iraq.[42] Al Qaeda in Iraq (AQI) sought to expel foreign interventionist forces and establish a symbolic Islamic caliphate, but their ignorance of local power structures and customs among their Sunni partners hindered that effort almost immediately. Most Sunni locals referred to AQI's operatives and soldiers as *takfir*, those who would use terrorism to further a violent extremist agenda against widely respected Koranic injunctions.[43]

Almost all insurgent attacks shared a common characteristic in that they utilized adaptive means to attack coalition forces. Insurgent groups usually operated in cellular networks, and generally did not openly attack the coalition's combat power with similar combat power. Instead, they employed techniques of attack such as Improvised Explosive Devices (IEDs), precision small-arms fire, Rocket-Propelled Grenade (RPG) ambushes, and "shoot-and-scoot" indirect fire attacks from mortars or improvised rocket munitions. Weapons for all of these styles of attack were plentiful from old Iraqi Army bases which were simply deserted or bypassed by coalition forces during the ground war. Gentile, then serving as the Executive Officer in the 1st BCT, 4th Infantry Division, recalled that upon the unit's arrival in Tikrit, they discovered thirty Iraqis openly looting weapons from an abandoned base.[44] In a 2009 interview, one sheikh from Al Anbar who eventually supported the coalition effort casually recounted taking 80 RPGs and additional PKC machine guns from an unsecured base after meeting the coalition forces and telling them about it the day before.[45] Insurgents posited that they could outlast a counterinsurgent effort by way of exhausting them with innovative attacks, and there was enough loose military material in Iraq to support this.

Since most insurgent groups used the same styles of attack and evasion, it caused initial confusion among American soldiers and a made many groups indistinguishable by their detected actions alone. This masked the underlying fact that while the insurgent groups were not actively supporting each other, they had progressed from an earlier stage of mutual antagonism.[46] Metaphorically, insurgent groups in Iraq during the early stages of the war were like card sharks working at the same table in Atlantic City: "we're not playing together. But then again, we're not playing against each other either. It's like the Nature Channel. You don't see piranhas eating each other, do you?"[47] However, familiarity breeds contempt, and the inherent divisions among these groups would become a key point of leverage for the counterinsurgency effort later in the war.

The Counterinsurgency Effort

The initial inability to contain the insurgency began within the context of the ground war in 2003. Several important aspects of the conventional campaign itself set the stage for the meager institutional efforts to improve civil capacity and pacification. The three main factors which created an environment conducive to an insurgency were: the coalition's flawed plans for reconstruction, the military's lack of training and preparedness for counterinsurgency warfare, and the effects of a De-Ba'athification policy.

A proper critique of post-invasion planning by the Bush administration and the American military is not that it lacked a plan, but that the plan lacked sufficient depth.[48] US House Representative Ike Skelton queried President Bush in a letter regarding plans for follow-on operations, asserting that it might be like a dog who chases cars finally catching one, and not knowing what to do with it.[49] Representative Skelton's wariness proved accurate, as the administration did not lack a plan as much as they lacked an appreciation for the robust requirements after the ground war. Wolfowitz is quoted as saying "I don't see why it would take more troops to occupy the country than to take down the regime," even with the benefit of historical data regarding the intense security requirements for the unrefined task of occupation.[50] Secretary of Defense Donald Rumsfeld drew on his previous experience with military officers in his attempt to keep overall force levels low, illustrating that "if you grind away at the military guys long enough, they will finally say, 'Screw it, I'll do the best with what I have.'"[51] From the outset, Franks' plan focused strictly on the amount of combat power required for the conventional fight, in an absence of the need for follow-on forces.[52] Operations after the ground war were supposed to include three foreign divisions, as well as assistance from allied paramilitaries such as the Italian Carbinieri to mentor Iraqi police forces.[53] One of the loudest dissenting opinions came from the Chief of Staff of the Army, General Eric Shinseki, who cautioned in his farewell speech to "beware the twelve-division plan for the ten-division army."[54]

Within the military, there was not an appreciation for these requirements either. Although some senior officers such as Shinseki identified the huge requirement for combat power and support units, there was not an institutional readiness for counterinsurgency warfare.[55] Hence, when the first signs of the insurgency emerged, there was not an inherent ability to defeat it at its weakest point. One Division Artillery commander recalls that "our biggest problem is that we didn't understand our responsibilities as an occupier; we just didn't. Your responsibilities are *everything*. And there was this false expectation that it would just fix itself."[56] Military preparedness and organization for civil support operations was so low that a division- or corps-sized unit could not be reasonably expected to run a state in America, let alone a province in Iraq.[57]

In 2003, the American military still considered the British Army to be the world-class force for waging warfare amongst the people.[58] British Army veterans of the campaign in Northern Ireland briefed possible insurgent tactics and appropriate responses at the small unit level, but American forces had already adopted nearly all of these by the summer of 2003.[59] At the BCT level, British NCOs and officers from the Operational Training and Advisory Group assisted in the development and training of framework operations within the unit's battlespace.[60] Although these early efforts were admirable, they could not overcome years of institutional neglect with both counterinsurgency doctrine and training in the American military. Due to the negative connotations from Vietnam that were associated with fighting an insurgency, operational reports in 2003 were deliberately void of the term and substituted phrases such as "former regime elements" to describe all insurgent activity.[61]

The third major factor that contributed to the coalition's inability to halt the nascent insurgency was the policy of De-Ba'athification. Against the counsel of many Iraqis, including the secular Shia interim Prime Minister Iyad Allawi, the Coalition Provisional Authority (CPA) issued a declaration that all Ba'ath Party members would be banned from holding senior officer positions in the army and government service jobs.[62] This effectively dissolved the army and almost all associated security forces, even though a later review of officer records indicated that only 8,000 of the 140,000 officers were committed Ba'athists. Collapsing the army also resulted in a predicted mass of 300,000 to 400,000 military-aged males.[63] Since Saddam Hussein's regime virtually required Ba'ath Party membership for all middle- and upper-class government jobs, there was a corresponding deficit in civil services throughout Iraq in 2003. This directly contributed to an initial environment without security in 2003, and lawlessness that the American military simply could not contain.[64] A British General Officer recalled a similar environment in Basra:

> I have no excuse for not understanding Mesopotamia intimately. When I arrived in central Basra, I was amazed by the scale of rioting and looting. It happens every time. I'd been to the Balkans and Kosovo. If you read any history book, and this was the third time the British have been in Iraq in 90 years, we had no excuse.[65]

The counterinsurgency effort began without institutional backing from the larger American military or diplomatic services. In 2003, stateside defense leaders directed the authors for the next National Military Strategy to remove all mentions of counterinsurgency from the document since they expected the campaign to be over militarily by the time the document circulated.[66] Throughout 2003, general officers edited operational assessments for terms and language before forwarding them to senior officials.[67] The provisional authority in Iraq, the Office for Reconstruction and Humanitarian Assistance (ORHA), was vastly undermanned and underfunded. Contrary to their original plan, one of Rumsfeld's senior aides visited Iraq and remarked to ORHA leaders that America did not owe Iraq the estimated $3 billion reconstruction effort.[68] A British evaluation judged that ORHA was "an unbelievable mess. No leadership, no strategy, no co-coordination, no structure and inaccessible to ordinary Iraqis."[69] With an absolute lack of assistance from ORHA, unit commanders had to implement governance, security, and development within the sectors they had seized at the conclusion of the ground war.[70]

The second iteration of provisional authority in Iraq was the aforementioned CPA, which was a result of the transformed and expanded ORHA, and headed by political appointee Paul Bremer.[71] Commanders in the field continued to be frustrated by the lack of progress as they waited for a centralized process to alleviate what they saw as a burden on military operations. Their assessments of CPA's effects at the street level ranged from "completely useless" to "ineptly organized and frequently incompetent," leading to the widely-held assumption that CPA really stood for Can't Produce Anything.[72] CPA focused on transitioning the central government functions to an Iraqi body as soon as possible, and Shia religious leaders to include Abdul Aziz Hakim and the Ayatollah Sistani pressured Bremer to hold direct elections instead of forming an appointed council.[73]

CPA's ineffective focus on large centralized improvements forced unit commanders to implement their own approaches. Major General David Petraeus' 101st Airborne Division (Air Assault) executed an operational approach of protecting the population center of Mosul, and implemented bottom-up governance solutions by cross-leveling the best practices of his battalion commanders.[74] The local population feared that the al-Jabouri tribe would fill the power vacuum left by the Ba'athist regime and the recently departed *peshmerga* force.[75] While Petraeus' unit conducted a form of counterinsurgency, most other units adopted an approach of "anti-insurgency."[76] Major General Raymond Odierno's 4th Infantry Division in the Salah-ad-Din Province illustrated this difference and adopted a tactic of large-scale clearance operations to detain suspected insurgents and their support networks. During the second half of 2003, their units in Tikrit experienced three times the attacks compared to forces in the rest of Iraq.[77] Simplistic analyses cast this as a "kill/capture" approach only focused on destroying the enemy's operating force, but Odierno's units faced a significantly different challenge than Petraeus' units in Mosul. In addition to the security and governance considerations, he had to relocate Kurds from Salah-ad-Din and Diyala provinces to northern enclaves, negotiate to preclude Iranian groups from interfering with reconstruction, and secure vast Iraqi military complexes in the Tigris River Valley.[78] Additionally, this was the heartland of the former regime, so the policy of De-Ba'athification had an extremely dispositive effect on the local population. A political adviser who had direct contact with both commanders during this time astutely criticized Ricks' narrative (as opposed to a historical account) in *Fiasco*, explaining that Ricks casts Petraeus as a savior and Odierno as a villain while counterinsurgency operations never have a clear distinction in disparate environments.[79] Although some heavy-handed tactics at the local level contributed to an antagonistic environment, Odierno's operational approach was simply a different technique to address security issues and the lack of effective governance.[80]

The lack of interunit synergy between military and CPA efforts was due to a lack of direction from the military headquarters, Coalition/Joint Task Force 7 (CJTF-7). Commanded by Lieutenant General Ricardo Sanchez, this headquarters was marred by a continuous disaccord with CPA, the absence of an effort to adapt operations, and the lack of a campaign plan.[81] Published reports attributed the differing concepts among division commanders as a direct result of a relative vacancy at the corps level.[82] Many of his contemporaries held the view that no military leader could have succeeded with the lack of institutional backing and resources that plagued Sanchez.[83] His own account lays blame at the feet of the US Central Command (CENTCOM), contending that they had a plan for the critical months of reconstruction after the ground war, but it was never promulgated.[84] All parties involved acknowledge a complete lack of coordination between Sanchez and Bremer.[85]

Coalition military forces operated with little information, witnessed by the 69 Human Intelligence (HUMINT) teams in Iraq only producing an average of only 30 reports per day in 2003.[86] Coupled with an extremely limited amount of troops, this led to unsophisticated and ineffective forms of security.[87] With this lack of security on the streets in most cities in Iraq, insurgent groups bought hired guns to fire at American forces for only $50 to $100 per attack.[88] The military used tactics such as massive detentions, H&I fires, and property destruction to combat this in the short-term, but they had damaging long-term effects in aggregate. By December 2003 it was clear that this form of security was marginally effective in securing American military forces, and even less effective in securing the population, evidenced by the fact that 62 percent of Iraqis cited security as their biggest problem. Additionally, 59 percent felt that the coalition were occupiers who would only leave Iraq by force.[89]

In early 2004, there were signs that the insurgent tactic of emplacing IEDs against patrols and convoys had a similar, if indirect effect. Major General Peter Chiarelli, commander of the 1st Cavalry Division, observed that "Coalition forces are forced to interact with the Iraqi populace from a defensive posture, effectively driving a psychological wedge between people and their protectors."[90] Due to the continued paucity of actionable HUMINT, units conducted large-scale sweeps and detained massive amounts of military aged males in restive areas. Some units such as the 82nd Airborne Division actively screened their detainees, only sending 700 of 3,800 detainees to facilities in Abu Ghraib during their deployment between August 2003 and March 2004. These measures were mildly effective, and internal estimates indicated that between 80 percent to 90 percent of all detainees in Abu Ghraib held no intelligence value in 2004.[91] As a reflection of these continued operations, Iraqi popular support for the CPA was down to 14 percent by March.[92]

These frustrations manifested themselves in the first two open uprisings against provisional authority in March and April, one in Fallujah and one in the Shia south. Sunnis in Fallujah felt particularly disenfranchised due to De-Ba'athification and the lack of continuity resulting from five different coalition units securing the area between the initial invasion and April 2004.[93] These tensions increased dramatically after a small arms engagement between paratroopers of the 82nd Airborne Division and two large demonstrations left 13 Iraqis dead and an additional 90 wounded in April 2003.[94] Immediately following a relief-in-place by 2nd Battalion, 1st Marine Regiment of the 1st Battalion, 505th Parachute Infantry Regiment, atrocities perpetrated against a contracted convoy generated intense American domestic political pressure to stabilize Fallujah. Marine leaders understood the consequences of sending a large conventional force to clear the city of insurgents, and decided to let the situation develop before appearing to attack out of revenge.[95] However, pressure from CJTF-7 prevailed and the Marines launched a large-scale combined arms offensive with inadequate provisions for civilians to flee the city. The large numbers of civilians present, a determined and growing enemy, and high volumes of firepower led to high levels of collateral damage. Combined with the absence of Iraqi security forces, the Arab media successfully portrayed this as an assault on Sunni society in Iraq, which attracted more fence-sitters to fight against the Marines in Fallujah.[96] After a failure to pacify the city with military force, the Marines presided over a unilateral cease-fire and transferred control of the city to the hastily formed Fallujah Brigade of Iraqi irregular security forces, which eventually disbanded in September.[97] When it became obvious that the Fallujah Brigade had no intention of integrating Fallujah into a wider Iraqi society in line with the CPA's goals, one Marine officer concluded that it was like using inmates to run the asylum.[98]

In the Shia strongholds of An Najaf, Kut, Karbala, and Sadr City there was a simultaneous yet unrelated uprising of Sadr's populist JAM force. Coalition forces underestimated both the loyalty and ability of JAM. Their response to the closure of Sadr's political newspaper and the detention of a close adviser caught many coalition forces by surprise. CJTF-7 ordered redeploying units to turn around in Kuwait and reinforce units in An Najaf and Karbala, while CPA sought a political settlement with Sadr. Although they would not admit it publicly, Sadr's forces were not required to surrender or disarm.[99] Unwittingly, Bremer and CPA had legitimized Sadr's political efforts by endowing him with artificial credibility for standing up to the occupying superpower and securing an advantageous outcome.[100]

Although British forces were able to maintain security in the furthest southern regions, attacks doubled countrywide during the spring and summer of 2004.[101] By alienating the Sunni community through the offensive in Fallujah and the Shia community through the confrontations with Sadr, the coalition was fighting a widening insurgency against both sects of the Arab population.[102] Coalition forces also saw issues with both sects' units in the fledgling Iraqi Security Forces (ISF). Sunni Iraqi National Guard (ING) battalions did not display the same competency as ING battalions derived from Shia militias and *peshmerga* units, and often flatly refused to fight against insurgent factions from predominately Sunni tribes. Additionally, there was still a number of militias and paramilitary forces that were not integrated into the ISF.[103] The predominately Shia National Police (NP) units which were first tested in Samarra took it as an opportunity to attack the Sunni population, a foreshadowing of events to follow in Baghdad years later.[104]

Some improvements within the counterinsurgency effort came at the end of 2004, beginning with the inevitable requirement to re-clear Fallujah with armed force. Politically, Fallujah had to be pacified prior to national elections which were planned for early 2005. To that end, American and Iraqi leaders exhausted all internal diplomatic efforts with Sunni leaders in the area before launching a largely conventional campaign to clear the city of as many as 6,000 insurgents. In contrast with the first effort in Fallujah, civilians were effectively evacuated and managed, and the US Department of State began building the municipal government.

This successful operation illustrated the effectiveness of recent changes in leadership and organizational structures. Allawi and the Interim Iraqi Government accepted sovereignty in June, allowing CPA to cease operations and transfer most of its functions to various echelons of military headquarters.[105] General George Casey took over as the top military officer in Iraq as the CJTF-7 headquarters transitioned to Multi-National Force–Iraq (MNF-I), and assembled a staff to analyze the effectiveness of the effort in Iraq. While this staff did bring a focus on counterinsurgency practices and issued a cogent campaign plan in August, the overall effort in Iraq still lacked an adequate strategic framework.[106] Retired Lieutenant Colonel Kalev Sepp developed an analysis of best practices in counterinsurgency which informed Casey's prescriptions for the conduct of operations.[107] Sepp represented Casey's educated, hand-picked advisers. In further contrast to Sanchez, Casey effectively worked with his diplomatic counterpart, Ambassador John Negroponte.[108] As a further improvement, Petraeus returned to Iraq to oversee the creation of ten permanent Iraqi Army (IA) Divisions as the commander of Multi-National Security Transition Command–Iraq (MNSTC-I).[109]

As with the transitions between Briggs and Templer in Malaya, or Westmoreland and Abrams in Vietnam, the temptation was to simply cast the successor as the savior of the operation. Casey's command improved the harmony of discrete operations in Iraq through steady improvement, but it did not significantly improve the operational approach or strategic framework. However, American forces began to sense some tactical victories which would carry them through 2005, instead of an unending string of local setbacks during Sanchez's command. Accordingly, Sanchez is viewed by many as the general who got it wrong instead of the general who was there at the wrong time.[110]

National elections provided a mixed measure of legitimacy for the central Government of Iraq (GoI) in January 2005.[111] Within this context, security operations in 2005 did not show much adaptation across the counterinsurgency effort as a whole, except for two critical operations in the remote western reaches of Iraq. Efforts to secure the towns of Tal Afar and Al Qaim in order to interdict Sunni *takfiri* infiltration routes from Syria and Jordan illustrated two successful tactical approaches, methods that would be adapted later to secure urban sectarian populations and reconcile to co-opt tribal forces. Both of these techniques would also directly affect the employment of artillery units since they relied more on the pragmatic use of dispersed ground security forces than a predilection for indirect fires.

When Colonel H. R. McMaster and the 3rd Armored Cavalry Regiment (ACR) arrived in Tal Afar in April, the town had been showcased by Sunni insurgents as their new centerpiece since losing control of Fallujah in November 2004. As such, they exerted de facto control over the mixed population.[112] The 3rd ACR employed a technique to sequentially separate and protect the population from the insurgency, clear the city of insurgents with judicious and limited use of firepower, and build outposts in the town.[113] To effectively separate the population from the insurgents and isolate the town for a high-intensity clearance, the unit shaped the edges of the town and allowed the population to move into prepared camps.[114]

One of 3rd ACR's advantages in this operation was its ability to redefine the struggle in the eyes of the population. Prior units in Tal Afar did not understand that the population had turned to AQI's false narrative of protecting them from other groups since there was no other alternative for security. As such, American forces conflated a cooperative population with the *takfiri* themselves before ceding control of Tal Afar.[115] The 3rd ACR was able to shift the narrative from Sunni versus Shia to Iraqi versus *takfiri* within the population.[116] The population came to see American forces as a relatively impartial guarantor of security, with many citizens moved to tears when the unit departed.[117] In the short term, the battle for Tal Afar severely disrupted AQI's ability to conduct unhindered operations in northern Iraq, and allowed the population to reject the *takfir's* brand of violence.[118] In the long term, the approach shaped how the Army eventually campaigned and fought successfully in Baghdad.

In Al Qaim, two Marine battalions cleared the border town of Sunni insurgents who had fled Fallujah in November 2004. The important phase of this campaign is what immediately followed.[119] The 3rd Battalion, 6th Marine Regiment, led by Lieutenant Colonel Dale Alford, engaged local Sunni tribes when they detected a rift between them and AQI forces in the area. Marine officers attribute this to their ability to understand the population in Al Qaim, due to their efforts to live amongst the people and conduct a census in the five local towns.[120] To exploit this opportunity, the Marines were able to reconcile with previously antagonistic Sunni tribes and integrate more than 700 men into new ISF units in Al Qaim.[121] The approach in Al Qaim influenced the Sunni tribes' response to AQI during the subsequent conflict in Al Anbar Province in late 2006, as well as the American forces' handling of the situation.[122]

Several improvements in the organization of the counterinsurgency effort complemented these two operations in 2005, but they were not enough to overcome a continuing focus on security in terms of enemy insurgents. Positive developments included a counterinsurgency center at Taji (the "Taji COIN Center") to immerse all incoming commanders in the successful tactics in use, tailored collective training for units prior to deployment, and specialized training for ISF advisory teams (called MiTTs).[123] Additionally, the Department of State replicated the Provincial Reconstruction Team (PRT) concept from the counterinsurgency effort in Afghanistan, with initial PRT members establishing contacts for future teams to be manned in 2006.[124] However, these advances could not make up for the continued effects of civilian casualties, property damage, and mass detentions in many areas. Overly centralized command structures and the continued ineffectiveness of new IA units exacerbated these flawed tactics.[125] As a reflection of many Iraqis' frustrations, an Oxford Research International poll showed that nearly 75 percent of Iraqis yearned for a return to self-determination under a strong central leader.[126]

The inadequacies of the counterinsurgency effort in 2005 led to increased domestic pressure against an open-ended military commitment. Units coalesced on progressively larger Forward Operating Bases (FOBs) while force protection concerns took on an increased primacy. President Bush outlined his plan to strategically disengage from Iraq, telling Americans that "[w]e will increasingly move out of Iraqi cities, reduce the number of bases from which we operate, and conduct fewer patrols and convoys."[127] Within Iraq, Casey expressed this operationally as an effort to stand down coalition forces while the ISF stood up, but for many units without capable partners this left a conceptual void.[128] A senior officer on Casey's staff recommended against expressing the act of transition as a separate line of effort in the campaign, contending that it should pervade all other efforts. After consideration, Casey decided that it would represent its own line of effort to emphasize the importance of transitioning to the ISF. This turned out to be a confusing artificial division for subordinate commanders and MiTTs.[129] Casey's dilemma was similar to the one facing Westmoreland in Vietnam; was the American counterinsurgency effort at a crossroads?[130] Some commanders sought exemptions to keep forces distributed in population centers or with ISF partners, since they cynically noted that they were "always six months from leaving Iraq."[131]

As the American-led coalition focused on the dual tasks of training the ISF and preparing areas for transition to full Iraqi control, the communal violence between Sunni and Shia elements threatened to pull the entire country into unabated chaos. A GoI referendum that allowed for limited federalism polarized many Sunnis and Shias over the future of Iraqi society and the division of natural resources. In February 2006, AQI bombed the Golden Mosque in Samarra, a holy site for Shia which irreversibly set sectarian warfare in motion in areas of mixed communities, chiefly Baghdad. The bombing achieved its intended effect, with JAM and the Badr Corps almost immediately beginning to cleanse mixed neighborhoods in Baghdad with lethal force.[132] Responding to JAM checkpoints and house-to-house clearing in the Jihad neighborhood of Baghdad, a Sunni member of parliament declared that "[t]his is a new step. A red line has been crossed. People have been killed in the streets; now they are killed inside their homes."[133]

Coalition assessments clearly showed that JAM had infiltrated nearly every facet of the ISF in Baghdad, meaning that supposedly legitimate forces were enabling or even carrying out much of the communal violence against Sunnis.[134] The Shias also used civic means of supposedly legitimate governance to disenfranchise the Sunnis. The Shia-controlled Ministry of Finance refused to open banks in Ameriyah, while 95 percent of banks in Shia neighborhoods were open. This meant that a Sunni had to travel through JAM checkpoints to get to a bank with the likelihood of a shake-down or even execution, or keep his money at home and run the risk of a shake-down at the hands of local AQI elements.[135]

This prompted a Sunni response in the form of AQI's vehicle-borne IED (VBIED) attacks against massed civilian targets in Sadr City.[136] Both sides settled into a steady dynamic cycle of AQI launching daytime VBIED attacks against Shia mosques and markets, with a response of JAM attacks at night against Sunni enclaves. Extremists targeted the opposite population, instead of attacking the opposite extremist group's means to inflict further atrocities.[137] Shia forces moved steadily across the northern neighborhoods of Baghdad in an attempt to exhaust the rapidly diminishing Sunni population with a mix of violence and controlled access to key commodities.[138]

At the time, the two pillars of the American strategic approach to counterinsurgency were a democratized GoI and a transfer to Iraqi security leadership; both were completely ineffective in addressing communal violence.[139] In an attempt to establish security within Baghdad through overt military presence, units partnered for Operations TOGETHER FOR-

WARD I and II. The plan called for partnered operations to clear extremist groups from the worst areas of the city, and the IA units to hold those areas. However, most IA units failed to show up, and the few that did were generally ineffective.[140] Iraqi police in Baghdad were almost completely ineffective since an estimated 70 percent of the force was affiliated with a militia group and involved in the violence.[141] By the completion of Operation TOGETHER FORWARD II, attacks actually increased by 22 percent and the coalition spokesman, Major General William Caldwell, admitted that the operation failed to meet its objectives.[142] Not only had Operations TOGETHER FORWARD I and II failed to curtail sectarian violence, they deprived other areas of critically required assets. One field-grade officer recalled that Multi-National Division–Baghdad (MND-B) was completely consumed in commanding and controlling so many units, and it gave the outlying BCTs nearly complete autonomy but few resources.[143]

Against the backdrop of this bleak environment in Baghdad, one promising development started in late 2006 in the Ramadi area. In what would come to be colloquially known as the *Al Anbar Awakening* or simply the *Awakening*, several Sunni tribes rejected the AQI narrative and actively stood up to the *takfir* narrative. They reconciled with American forces and ISF in the area, and eventually integrated into the security structure in order to protect the local population.

By late 2005, AQI had already began to wear out its welcome with local Sunni tribes by riding roughshod over tribal rights and customs, to include the takeover of lucrative smuggling routes from the west.[144] One of the first tribes to actively oppose AQI, the Albu Issa, took its cue from the earlier operations against AQI in Al Qaim.[145] Coincidentally, the new unit in Ramadi (the 1st BCT, 1st Armored Division led by Colonel Sean MacFarland) took an inventive operational approach to the perceived lack of a coherent method from earlier units.[146] Contrasting the efforts to disengage and transfer control in other areas, MacFarland visualized a plan to isolate insurgents and deny them sanctuary, build the ISF, clear and build combat outposts among the population, and engage local leaders to determine which ones had the most respect among the tribes.[147]

As tribal elements provided manpower for Iraqi police stations, they could position the first 100 in a substation of their own preferred location.[148] With this ability to generate combat power through the existing framework of the ISF, they eventually became the main effort and the focus changed from clearing AQI from Ramadi to sealing them off inside the city for easy defeat.[149] Overall, the successful integration of tribal forces into security framework operations in Ramadi proved that Iraqis could target extremist groups, remain armed, and not necessarily descend into chaos.[150]

The operations in Ramadi also illustrate that many American units were beginning to adapt to the requirements of counterinsurgency operations. Units at the battalion level and below were generally internalizing lessons and making astute *ad hoc* arrangements, but no unifying themes or practices abounded across Iraq.[151] Many small-unit leaders began to self-educate, and the overall trend was improvement through bottom-up refinements.[152] Units began to fight the IED threat holistically, nurture and integrate HUMINT systems, and establish outlying patrol bases when they could avoid the demands from higher echelons to consolidate on larger FOBs.[153] Petraeus made an effort to establish the missing linkage of unifying themes through an improved counterinsurgency doctrine, manifested in a large conference at Fort Leavenworth to discuss a rewriting of existing Army doctrine.[154]

As American forces made progress in 2006, the ISF remained a problematic institution. The IA showed the most improvement, due in part to their relative longevity and growing base of experience.[155] The NP units were significantly more of a hindrance to the counterinsurgency effort, instead of the critical paramilitary force that should have been providing security within the urban areas of Iraq. In most cases, they were equally as destabilizing as the insurgents themselves since they engaged in the sectarian violence against Sunnis in Baghdad from a position of theoretical legitimacy.[156] The Iraqi Police had an uneven quality from city to city, but in general they were vastly undermanned, underequipped, and able to do little besides maintain marginal control at traffic circles and checkpoints.[157]

Against the evidence of the ISF's inability to provide effective security without significant coalition assistance and advice, the strategic focus on Casey's staff was still disengagement via transition throughout 2006.[158] Although the security situation seemed to be getting progressively worse by the week, CENTCOM continued to cancel BCT deployments or hold them in Kuwait for further disposition. According to a strategic plans officer at the time, the CENTCOM planning assumption was that the US military was in a lockstep march to reduce from twenty BCTs to ten BCTs by the end of 2006.[159] Indeed, the situation was getting worse by the week, as sectarian killings in Baghdad alone reached 125 per night.[160]

The violence of 2006 forced a review of the national strategy regarding the continuing counterinsurgency effort in Iraq. A congressionally mandated Iraq Study Group released its findings in December, which recommended a continued disengagement from Iraq. To achieve this, the group's report advocated two simultaneous approaches to achieve a satisfactory strategic outcome and a complete withdrawal of the American military.[161] This prompted the American Enterprise Institute (AEI) think tank to develop an alternate option for counterinsurgency in Iraq, focusing on securing the population within Baghdad which would provide breathing room for GoI to develop. The Executive Summary for the plan began with the assertion that "[v]ictory is still an option in Iraq."[162] Simultaneously, Odierno returned to Iraq as the commander of Multi-National

Corps-Iraq (MNC-I), the operational military headquarters directly subordinate to Casey. Odierno reversed the military aspects of the strategy to extricate coalition forces from urban areas and into large FOBs, in clear opposition to Casey's plan.[163] Although American forces steadily adapted to achieve a very real tactical capability in counterinsurgency by 2007, these revised strategic objectives and operational plans would be what came to be known colloquially as *The Surge*.

President Bush enacted the AEI plan and announced the strategy on 10 January 2007, against the advice of Generals Abizaid, Casey, and the JCS, who all viewed coalition military presence as the primary driver for instability in Iraq.[164] At the same time, President Bush replaced Casey with Petraeus to implement the new strategy at MNF-I. While Petraeus' educated staff and leadership undoubtedly influenced the successful transfer of lessons learned from successful operations in Mosul, Tal Afar, Al Qaim, and Ramadi, strategic planners at the time credit Odierno's operational concept as expressed in the Baghdad Security Plan as the key improvement.[165] Odierno's inbrief to Petraeus in February presented the key tasks of securing the Iraqi people in Baghdad and interdicting the accelerants of sectarian violence in the Baghdad belts, with the neutralization of the actual extremist networks supplementing these two tasks.[166]

The effects of unrelenting sectarian violence in Baghdad made it difficult to measure the actual effectiveness of additional combat power, a refined operational approach, and an improved strategic framework. By the time additional BCTs arrived in Baghdad, ethnic cleansing was largely accomplished in formerly mixed neighborhoods so violence was about to trend down naturally. As one intelligence officer remarked, "[n]ow that the Sunnis are gone, murders have dropped off . . . one way to put it is they ran out of people to kill."[167] A civilian adviser to MNF-I and MNC-I took this analysis a step further, asserting that the Shia had already "won" the civil war and Sunnis were left to wonder what place they had at all in a new Iraq.[168] Within this context, remaining Sunnis in Baghdad relied more than ever on AQI and other *takfiri* elements for security in their enclaves.[169] Sectarian cleansing in Sunni neighborhoods continued as the initial American BCT entered northwest Baghdad; a Shia IA battalion forcibly evacuated approximately 50,000 Sunnis from the Hurriyah neighborhood alone.[170]

The Baghdad security plan, Operation *Fard al-Qanun*, provided a new articulation of the clear–hold–build methodology. This plan also framed operations in a context that truly integrated the ISF leaders at all echelons. Successful operations featured a combination of partnered Joint Security Stations (JSSs), the compartmentalization of sectarian populations, and the integration of irregular security forces. Across Baghdad, American forces deployed among the population to establish 32 JSSs in more locations than originally envisioned.[171] JSSs represented the tactical efforts of *The Surge*, enabling partnered operations among the population with a decreased reliance on standoff for force protection and other risk-averse measures. This required a shift in soldier's attitudes, since they departed from the relative comfort and security of the large FOBs.[172] As Sepp explained in his analysis of operations in Tal Afar:

> If you really want to reduce your casualties, go back to Fort Riley; it's absurd to think that you can protect the population from armed insurgents without putting your men's lives at risk. [Gathering troops at enormous bases] is old Army thinking—centralization of resources, of people, of control. Counterinsurgency requires decentralization.[173]

JSSs endowed small unit leaders with a truly partnered ISF force and the ability to share intelligence and analysis capabilities. Company commanders learned that the ISF's informal HUMINT networks generally outperformed American HUMINT teams for local atmospherics and background information, and often the ISF commander's cell phone was the lynchpin to successful partnered operations.[174] Additional clarity on the roles of MiTT elements allowed them to function as advisers to the ISF commanders and staffs, with tactical units allowed to train and operate directly with the ISF unit itself.[175] The chief limits on these partnered operations were the immature ISF logistics system and the misaligned boundaries between ISF and American forces. Although company-level commanders sought to bolster their ISF counterparts' initiative while responding to crises, some of MND-B's artificial boundaries precluded this until they were re-aligned.[176]

JSSs allowed commanders to penetrate the population in order to understand the environment and establish security against extremism and continued communal violence. Since the Baghdad population was exhausted by continual violence, they were willing to support whomever appeared strong enough to win and offer them impartial security. When American forces moved out from distant FOBs, it projected this strength to defeat extremists at the local level.[177] Astute company-level commanders also aligned local governance with the security inherent in JSSs, creating the ability to leverage combined efforts of the ISF, the local economy, government officials, and tribal leaders.[178]

Another successful feature of the Baghdad security plan was the compartmentalization of sectarian communities. Through the means of concrete barrier walls and checkpoints, partnered forces sought to control the environment and keep the accelerants of sectarian violence separate from the population, rather than control the population itself. One battalion commander illustrated the disruptive effect of this effort:

> Baghdad has a big highway network and the insurgents exploited that and were able to move much more quickly than we were, virtually undetected. By putting up the barriers we took

away the advantage of mobility and their ability to move undetected because they had to move through checkpoints. Even if it was a shoddy checkpoint run by IA, the fact that it was there was enough of a deterrent.[179]

This effect was evident almost immediately, as AQI continually attempted to establish holes in the barrier walls while security forces maintained them.[180] Some American forces implemented adaptive approaches to compartmentalization, such as installing swing-arm gates that removed the stigma of continuous concrete barriers but were still effective in denying access to VBIEDs and massed insurgent movements.[181] Since Baghdad neighborhoods were relatively homogeneous by the time additional forces arrived, the population could be compartmentalized along approximate sectarian fault lines, something that would not have been possible earlier.[182]

A third aspect of the improved counterinsurgency approach was a sustained effort to recruit, co-opt and integrate local security forces into the framework of population security.[183] A battalion commander, who served as an operations officer at the time, noted that it was like a wave of Sunni moderation emanating from Al Anbar that local commanders could exploit.[184] Petraeus' contributions to this effort were the recognition of its importance and utility, and nationalizing the program to portray the Awakening as a turning point in the war.[185] Efforts to establish local security forces was a mix of seeking reconcilable forces and supporting capable forces as they self-identified, and a spike in the price of black market AK-47 rifles showed evidence that local groups were rushing to arm themselves as potential candidates in some areas.[186] Working with these irregular local security forces was a challenge which required intrepid commanders who could visualize and integrate these forces to complement ISF operations, as illustrated in the following description:

> Indigenous forces have a lot of latitude that we don't have, they were not inhibited by ROE the way we were. Its rough justice . . . it's the messy and dark side of working with indigenous guys. You have to understand it and be willing to accept that. If you can live with that, and I can, then you're fine. If you're trying to change their culture and their way of war to be our way of war then you'll be there a hell of a long time.[187]

The effect of these forces was almost completely positive, in contrast to some early concerns that arming militia forces outside of the reach of GoI would send Iraq into further chaos or that American leaders were simply bribing insurgents not to attack them.[188] This was due in large part to the small unit commanders' adaptation to view them as partners, not employees or former insurgents.[189] In a parallel to the expanded activity with ISF units, partnership bred an exchange of ideas and HUMINT that allowed coalition forces to improve their targeted attacks on violent extremist cells. Paradoxically, a focus on securing the population through an indirect approach yielded an expanded capability to directly attack the enemy.[190] The effect within the local population was generally outstanding, since the irregular forces' wages usually remained in the local economy and the neighborhood could take pride in defending itself.[191] In one BCT commander's area in Baghdad alone, they went from finding an average of 500 executed bodies per month to a pacified environment in which locals brought American soldiers food on holidays.[192]

The glaring negative effect was that most of these forces were Sunni since the Shia community enjoyed better protection from the JAM-infiltrated ISF in Baghdad. As such, they would not truly reconcile with the Shia-dominated GoI and demanded eventual integration through ISF or public sector jobs which simply did not exist. As one civilian adviser to MNC-I noted, "20 percent were supposed to go to security. But then, for the government, why would it give jobs to bad Sunni boys before good Shia boys? How would it then manage its constituents?"[193] The GoI bought into reconciliation for most Shia groups, but viewed Sunni groups as irreconcilable.[194]

As discussed above, actual effects of the improved counterinsurgency efforts in Baghdad of 2007 are difficult to isolate due to the exhausted nature of the population when it began. Additionally, coalition forces gained security block by block in Baghdad, where a soldier could walk the streets without body armor in one neighborhood less than a quarter mile from intense urban combat in another.[195] Some violence also migrated into the Baghdad belts and further to the Salah-ad-Din, Diyala, and At Tamin provinces.[196] The captured diary of the regional AQI commander in late 2007 showed that reconciliation in Salah-ad-Din robbed Jaysh al-Islami and AQI of combat power. He lamented that "there were over 600 fighters in our sector before the tribes changed course" but he only had twenty reliable fighters remaining after coalition forces adapted to integrate irregular tribal security in the area.[197] Al Anbar continued to reap the benefits of pacification from late 2006, and by the summertime American forces were able to stage a symbolic fitness run along a previously deadly stretch of highway that had claimed two soldiers in an IED attack just six months earlier.[198] Due to the increases in counterinsurgency proficiency and newly reconciled tribal forces, Sunni areas just outside of the belts such as Ramadi, Hit, Tikrit and Fallujah saw civilian death rates drop precipitously.[199]

Insurgent tactics within Baghdad adapted to the new environment, and attacks against American forces were almost exclusively Explosively-Formed Projectile IEDs (EFPs) and precision small arms fire attacks.[200] While some units remained committed to dismounted patrols, most used increased amounts of protection and standoff to mitigate these risks. This served to further isolate American soldiers from the population while on patrol, and only marginally increased survivability.[201] The domestic media's perception of a rise in casualties during *The Surge* is most likely a result of increased direct contact with insurgent forces in a predominately urban environment.[202] However, there is some validity to the notion that many extremist elements felt the urgency to "use it or lose it" with respect to weapons caches and combat power in Baghdad.[203]

The effect on JAM extremists was catastrophic, as Sadr prudently chose to use the new environment to streamline his militia and political organization. As JAM lost its sense of coherence and discipline, Sadr emplaced a cease-fire and attempted to integrate into national political efforts on his terms. To address the rogue elements of JAM, known as the Special Groups, he formed the Golden Battalion. Petraeus met with moderate Shia leaders to exploit this schism as a route to increased security.[204] In Kilcullen's assessment, Sadr and his closest allies made an attempt to eliminate JAM's violent elements "[b]ecause we treated them as the authority."[205]

Within the American military, the effects of this successful counterinsurgency approach took root at both institutional and small unit levels. Institutionally, the considerable yet reversible security gains in Baghdad granted Petraeus the political capital to develop a strategy to assist the GoI in consolidating these gains. With the rediscovered cachet of counterinsurgency, he and his supporters had increased influence within the Pentagon to make this happen.[206] In recognition of the effectiveness of incoming units' education at the Taji COIN Center, British forces sent elements of their next incoming brigade there in February 2008.[207] The largest institutional effect on the military was a resulting examination of the counterinsurgency doctrine encapsulated in Field Manual 3-24, *Counterinsurgency*. Many practitioners in Iraq held the view that the new manual simply distilled practices and concepts that were already in action when it was published in 2006. This notion prompted the Taji COIN Center to tell at least one BCT that they did not need to worry about an impact of the new text since they were already adapting, and all the manual did was codify the adaptation.[208] The success of the new counterinsurgency approach also prompted an institutional discussion on the merits of a population-centric approach compared to an enemy-centric approach to modern counterinsurgency warfare. Although many leaders felt that having an "intellectual knife fight over [the proper model] is healthy" to flesh out the nuances in counterinsurgency, they identified that an approach must be pragmatic and suited to the specific environment.[209] Additionally, the doctrine did not acknowledge that sometimes a counterinsurgent force will have to fight rogue elements of the host nation government which stretch the limits of legitimacy.[210] Other officers simply felt that the discussion engendered a false dichotomy.[211]

At the small unit level, the most significant effect was the increased reliance on junior leaders to operate in a decentralized role. Military commanders internalized the lesson that empowering junior officers allowed a unit to account for a greater level of detail in counterinsurgency operations.[212] Commanders also lauded the effectiveness of embedded PRTs if they were properly resourced and secured.[213]

As operations in Baghdad and the belt areas leveled off in late 2007, the focus turned to consolidating operational gains and enhancing the effectiveness of the GoI. To this end, reconciliation of tribal elements and rehabilitation of former insurgents gained importance. By focusing on reintegrating former insurgents instead of simply releasing them, coalition forces supported existing power structures in Iraqi society. Some units were able to release detainees to tribal leaders who would take responsibility for them, allowing the individual to retain his honor.[214] The inevitable transfer of authority to the GoI for detainee operations forced the coalition to make a clear distinction which insurgents could never be reconciled and which ones could be rehabilitated.[215] The combination of a command climate conducive to empowered junior leaders and a necessity for reconciliation meant that company-level commanders could usually manage the complex relationships to co-opt existing groups and leaders within their areas of operation. One commander described using his knowledge of a mayor's nefarious dealings as leverage to indirectly control his area of operations since this the mayor had a strong influence on the levels of violence. This commander felt that his method was better than having 150 soldiers, and more effective than any bomb he could drop.[216] In an adjacent area, another commander successfully used his understanding of the local power structures to co-opt an insurgent leader and hand him limited responsibility for security along a critical stretch of highway that criminally corrupt IPs had all but abandoned.[217] That two company-level commanders would be permitted to liaise with insurgents or their direct enablers would be nearly unthinkable to most units just a year prior.

In 2008, it became apparent to outside observers that the British counterinsurgency effort in Basra was not experiencing the same gains. British armored units had already cleared Basra once during the restive summer of 2004 amidst the general Sadrist uprising, and their application of tough but judicious force was supported by most Iraqis in the region.[218] MNF-I had always left Basra to British land forces, and commanders in MND-SE enjoyed a considerable latitude which, in retrospect, some senior British officers deemed too great.[219] The lack of a viable ISF presence in Basra greatly hindered operations since the British did not advise their partnered units' staffs and commanders with the MiTT model, and the IA Division responsible for Basra was the lowest priority for resources within the Iraqi Ministry of Defence.[220] British forces had a low level of confidence in the prospect of an American-led coalition succeeding in 2007, and the UK began to look for an exit from Basra that did not look like a defeat.[221] This method had historic precedents, since British forces withdrew from Palestine in 1947 and Aden in 1967 under political pressure rather than confront insurgent violence directly.[222]

In late 2007 JAM militias increased their hold on Basra and inflicted casualties against the British force, which increased domestic political pressure for a withdrawal.[223] The British did not have sufficient combat power, nor political will, to clear Basra in a refined manner similar to Tal Afar in 2005, or even in a heavily conventional campaign such as Fallujah in 2004.[224] Ahmed al-Fartusi, a detained JAM leader, claimed that he could negotiate a deal with JAM leadership in Basra to end attacks against the British if they released JAM detainees and withdrew from the city onto their large FOB outside the urban area. This was pursued in what is now known as *The Accommodation*, sold with the tagline of "it's Palermo, not Beirut" in an attempt to cast the violence in Basra as a criminal matter instead of an insurgent matter.[225] All other British intelligence-gathering platforms were sent to Afghanistan, and given the lack of advisers with ISF units it meant that the MND-SE headquarters "had lost the situational awareness on which the policy of overwatch had been predicated."[226]

Rocket and mortar attacks resumed against British forces equal to the conditions before *The Accommodation*.[227] Due to the political sensitivities to reopening a violent conflict, counterbattery fires at these targets was extremely constrained: "to fire back on the [indirect fire] we had to get authority from the Prime Minister. Signed off from the Prime Minister, with a collateral damage report going back. So, it was quite a big thing for us to fire back."[228] With what he perceived as minimal British support forthcoming, Mohan flew to Baghdad to brief Odierno, Petraeus, and Maliki on a deliberate counterinsurgency plan to re-take Basra. After Odierno told him that there were not enough military resources available and the offensive would have to be sequenced later in the year, Maliki engaged in brinksmanship and ordered the mission to proceed immediately knowing that Petraeus would support him. Maliki understood that by making it a test of GoI resolve, Petraeus could not allow it to fail due to lack of American support.[229]

The resulting surge of resources and ISF units with American MiTT elements made the difference in heavy fighting to recapture the city.[230] American mentors to the highest ISF echelons monitored the deployment of plainclothes Iraqi soldiers in Basra to warn the population, coordinate with them, and try to gain support for Operation CHARGE OF THE KNIGHTS. Without overt measures for population relocation in Iraq's second largest city they expected a catastrophe, but it didn't happen.[231] Iraqi members of parliament and Iranian Quds Force operatives brokered a cease-fire in Basra, which abruptly ended partnered ISF and British combat operations well short of the estimated month-long duration.[232] Operation CHARGE OF THE KNIGHTS reaped a huge psychological effect for Maliki and the GoI, showing that they were effective and a loyal ISF was in charge of the city. Additionally, the security gains in Basra allowed the

British military to change their approach with ISF mentors and establish an effective counterinsurgency framework, meaning Basra did not end in a British defeat.[233]

There were several institutional factors that contributed to the crisis in Basra. Primarily, there was not a strong level of introspection and analysis fostered by leaders in MND-SE. This stands in contrast to leaders such as McMaster, Alford, and MacFarland who fostered pragmatic approaches to amorphous challenges in Tal Afar, Al Qaim and Ramadi respectively.[234] The British Army also operated in a very limited area in the southeastern region of Iraq with a relatively small force, so it did not have as much contemporary institutional experience building up during OIF as the American military.[235] One of the largest issues was the operational and strategic disconnect with the MNC-I headquarters; however, this was a symptom to a larger issue. An American planner in Baghdad judged that the British Army acted as if Petraeus and Odierno had never come to Iraq and continued to follow Casey's approach of disengagement and transition.[236] There were also challenges in how the British national command element nested within a theater coalition headquarters. The new British Ministry of Defence command structure for overseas operations was also relatively new, and did not have significant organizational experience balancing political and military considerations in operations larger than the Balkans and Sierra Leone.[237] Regardless of political constraints, the British contingent in Basra did not account for the changing security environment in 2007 and 2008.[238]

The ability to dynamically re-task critical assets (both Iraqi and MNF-I) illustrated the improved situation across Iraq in 2008. The security situation and a stabilizing GoI gave Maliki the confidence to address Shia violence in Basra and Sadr City in March. The Sadr City violence also slowed talk of transition during a CENTCOM review of the strategy, and gave the counterinsurgency effort a slower and sustainable pace. As leaders began talks of transition and a new Status of Forces Agreement between the GoI and American governments, Maliki would often joke to Petraeus that it seemed the Americans wanted to stay in Iraq more than the Iraqis did.[239] In September, Odierno was promoted to replace Petraeus as the MNF-I commander, providing crucial continuity to the counterinsurgency effort.

One of the largest issues facing commanders at the end of 2008 was the effort to integrate the irregular local security groups that were so critical to security gains during the previous two years. The program became known as the Sons of Iraq (SOI) in a coordinated effort to integrate the most capable fighters into the ISF or public sector jobs. By this time, most ISF units were accustomed to the SOI presence at checkpoints and willingly integrated them into security framework operations at the local level. SOI groups were effective in securing limited local areas through checkpoints and an intimate knowledge of the neighborhood population and power structure. Continuing into 2009, some unit commanders struggled through a generally inefficient program to integrate individuals, but saw instances of success when specific SOI members were fully trained Iraqi policemen or soldiers in their area.[240] Difficulties were generally due to inefficient pay systems, the aforementioned lack of jobs available in the Iraqi public sector, and the fact that some groups cooperated with American forces instead of the GoI.[241] One civilian adviser saw this effort as proof that the underlying sectarian divide still remained as strong as ever, since real reconciliation would involve admitting to former acts of insurgency and terrorism. This arrangement was more akin to a series of cease-fires rather than true reconciliation and integration of Sunni and Shia militias.[242]

By late 2008, the increased operations and a refined approach in Baghdad was seen as a success by most observers in Iraq. This provided the dual opportunities to rehabilitate Baghdad and transition responsibilities to ISF partners. Odierno's director of strategic operations, Major General Guy Swan, assessed that "with the security gains, there is a window of opportunity . . . only they can do it. We have set the conditions for them. They have an opportunity to pursue their own destiny."[243]

Rehabilitating Baghdad and the belt areas after years of communal violence and militarized security was an effort that the population supported. One battalion commander recalled citizens bringing concertina wire to one of his JSSs north of Baghdad, and that they were genuinely happy to see the vestiges of a combat environment removed from their neighborhoods.[244] Commanders used their discretionary funds to focus on projects that would create a local perception of a return to normalcy and stimulate the economy.[245] Requirements for intelligence at the small-unit level in Baghdad were no longer focused on local thugs, but on local leaders with influence in economic and social programs. One battalion commander southeast of Baghdad concluded that his most important information was what he could use to inform embedded PRT operations, and JAM's influence was almost irrelevant by that point.[246] Partnered offensive efforts to target AQI and JAM Special Groups leaders and their networks became almost a nighttime caveat to daytime civil support operations in the minds of many company-level commanders.[247]

Transitioning security responsibilities to the ISF in these areas became another focus, with the Federal Police (formerly the National Police) divisions assuming responsibility in Baghdad in July of 2009. American forces in urban areas concentrated efforts on a responsible withdrawal to hand over JSSs and combat outposts to direct ISF control, while they would continue to advise operations and secure the 23 PRTs operating throughout Iraq.[248] After the formal security agreement took effect on 30 June 2009, tactical units did not strictly follow the blanket policy of American combat forces' withdrawal from urban areas, which led to several confusing corollaries about what constituted a combat patrol and a general disconnect between headquarters at different levels.[249] As an unintended benefit, this enabled ISF growth at higher echelons more than almost any other initiative in the counterinsurgency effort, since MiTTs and partnered commanders could truly only advise operations instead of lead them.[250]

By the end of 2009, reported civilian deaths were below pre-invasion levels.[251] As BCTs redeployed they were replaced by Advise and Assist Brigades (AABs) to maintain a focus on improving the quality of ISF units instead of continued operations against the remnants of JAM and AQI. Eventually, enough BCTs redeployed without replacement that one AAB had responsibility for all operations in Baghdad.[252] Both Iraqi and American leaders planned for a conditions-based withdrawal from Iraq to avoid mistakes of previous counterinsurgency campaigns.[253] Iraqi leaders felt that continued mentorship to develop the GoI's capacity for governance was needed and that "[t]he withdrawal of U.S. forces from Iraq should mark an evolution in the U.S.-Iraqi relationship, not Washington's disengagement."[254] To this end, Secretary of Defense Robert Gates declared that OIF would transition to Operation NEW DAWN (OND) beginning in September 2010.[255] Concurrently, MNF-I transformed to US Forces–Iraq which "advises, trains, assists, and equips Iraqi Security Forces, enabling them to provide for internal security while building a foundation capability to defend against external threats."[256]

OIF provides a case study in success through tactical and operational adaptability, with several potential long-term effects. In a return to Baghdad in December 2010, a former BCT commander saw incredible improvement at the street level. Only some of the concrete barrier walls remained, violence was low, and the bustling population was building new markets.[257] This scene was an unlikely image in early 2007, when Petraeus arrived to see the bombed-out "ghost towns" of former Sunni enclaves in Baghdad.[258] However, tactical and operational successes must be tempered with enduring strategic issues. The US (especially the Bush administration) was plagued by the lack of a comprehensive WMD network or evidence that AQ was intimately involved with Hussein's regime. AQ's actions in fighting and damaging American prestige served to improve its narrative among radical Islamists. A civilian adviser to MNF-I summarized this mixed result:

> What it has done to our reputation . . . the day after 9/11 you had support across the world. People in Gaza, people in Iran demonstrating and raising money for America. And you look at how we went about responding; how we went about trying to make ourselves safer, how many more grievances we've created along the way. We killed tens of thousands of people, we've got more enemies today than we've ever had before. It's a bit like "Team America." It's funny but it's so realistic. And now we understand how to do COIN, but we're not going to be

allowed out again.²⁵⁹

The Employment of Artillery Units

The employment of artillery units within the counterinsurgency effort began during the consolidation after the ground war, which is notable because American forces leveraged the lowest ratio of artillery to maneuver forces since the Spanish-American War.²⁶⁰ Artillerymen's contributions in the war took many forms to include the delivery of indirect fires, security as converted maneuver forces, advisory and training of ISF units, and a host of other missions inherent to a counterinsurgency effort.

In some areas, artillery units began the counterinsurgency campaigns of OIF much in the same way they spent most of the counterinsurgency campaign in Vietnam: delivering H&I fires within an ineffective framework of operations. Odierno's 4th Infantry Division in Tikrit used H&I fires against suspected enemy locations in what he referred to as proactive counterfire to mitigate enemy rocket and mortar threats against the ever-expanding FOB locations.²⁶¹ Farther south in the division's area of operations in Baqubah, units fired more than 200 rounds of artillery and heavy mortars in July 2003 alone. While this number pales in comparison to the amount of munitions used in Vietnam and even Malaya, it is indicative of the continued employment of artillery units in the indirect fire role during the incipient stages of the insurgency.²⁶² However, commanders began a steady habit of employing artillery units in other roles as the indirect fires infrastructure matured in OIF.²⁶³

Vast improvements in artillery range and accuracy impacted the ability to deliver indirect fires in a counterinsurgency; concurrent improvements in observer equipment greatly minimized both observer location and target location errors from the days of using orienting rounds in Vietnam. Within this environment, the new precision Guided MLRS (GMLRS) munition became a weapon of choice for many fire supporters and their commanders after its introduction in Iraq in 2005. Due to its precision, range, and relatively lower collateral damage, the GMLRS enabled coverage and responsiveness similar to the use of distributed FSBs in Vietnam but without the intense resource cost. This also restored the conceptual dynamic of mixed general support and direct support artilleries available to a commander for specific operations, with the option to re-task direct support artillery in a nonstandard role. Since the volume of indirect fires was so low compared to Vietnam, the practice of only using artillery as an emergency measure for troops in contact and selected missions made the indirect fires architecture similar to Malaya. However, the extended range of GMLRS munitions (70 kilometers) and 155mm howitzer systems meant that artillery units did not have to break into smaller troops as the RA did in Malaya, and they could co-locate on large FOBs instead of requiring infantry and helicopter support for FSBs in Vietnam.²⁶⁴

As the indirect fires infrastructure matured in OIF, it led some leaders to question the need for artillery fires at all. Major General James Mattis decided that his 1st Marine Division would not bring any artillery or tanks when they replaced the 82nd Airborne Division in Fallujah in 2004. He explained his approach to insurgents this way: "I didn't bring artillery. But I'm pleading with you, with tears in my eyes: if you f--- with me, I'll kill you all."[265] This approach seems to conflate the lethal potential of an artillery unit's indirect fires with collateral damage, and does little to mitigate risk against a severe increase in the scale of attacks. After experiencing the indirect fire threat to locations around Fallujah first-hand, Swannack was able to prevail on him to change: "[a]fter seeing how we got mortared and rocketed in the evenings, they decided to bring it."[266] That proved to be a pivotal decision for Mattis, since Marine artillerymen fired more than 4,000 artillery rounds and additional 10,000 mortar rounds during the second attempt to clear Fallujah in 2004.[267]

After Fallujah, many commanders and local leaders sought to conduct operations with minimal collateral damage. Inevitably, this was rooted in a mix of perception and reality but the effect was a decreased use of artillery indirect fires in general. A former mayor of a prominent Iraqi city with military experience illustrated this wariness as coalition forces prepared to clear his city of insurgents in 2005:

> I refuse any fighter jets to participate in the operation. And I refuse any artillery to participate. If the fighter jets participate, the first bomb that drops, I will resign. Because that is what happened in Fallujah and was a source of a lot of problems. They said, "How can we support our troops then?" I said Saddam Hussein in Basra had no air support when he cleared it out. Why don't you just rely on helicopters and mortars?[268]

This decreased use did not completely preclude the use of artillery fires, and several refined techniques emerged to capitalize on the benefits of artillery for counterfire and observed H&I fires.

The use of artillery units' indirect fires in Ramadi in 2005 show a refined integration of fires and effects with the commander's vision. The BCT's artillery battalion fired thousands of fire missions for terrain denial and illumination, and the BCT commander lauded these missions as a "Ph.D.-level counterfire" method which deprived the enemy of the ability to adjust fire on coalition FOBs.[269] The key to this effort was not so much the refined techniques, but that the commander understood the psychological effects of using artillery in the outskirts of Ramadi and adjusted it accordingly.[270] Other commanders also understood the use of artillery fires to interdict or disrupt insurgent activity as an effect to enable larger operations.[271] As with Malaya and Vietnam, commanders rediscovered the ability to use indirect fires to flush insurgents into advantageous terrain: "chasing artillery with artillery will get you nowhere. And we practiced the terrain denial piece to force them into areas where we could engage them. That was the key."[272] Still, some units used ineffective and unrefined techniques of employing artillery firepower to deny terrain and disrupt enemy actions. One battalion commander saw this firsthand, as he critiqued the BCT's use of artillery in his area:

> Then they were shooting terrain denial fires. They would shoot based on this information, so they would just shoot in these fields. And we were just churning up these fields. The AQ were going around saying that the Americans were scared like women. The other line was that we were there but we are invisible and magical and just can't be hurt. So the question is: what are we accomplishing here?[273]

By the time MNC-I arrayed the additional BCTs in Baghdad in 2007, that large urban area was almost completely devoid of indirect fires coverage. One BCT commander was told flatly that he could not use his own direct support artillery battalion to provide fires initially, and in his estimation "people had forgotten that we could use control measures to do this safely."[274] Eventually the requirements for fires in Baghdad trumped the institutional notion that collateral damage was unavoidable and unacceptable. BCTs were able to leverage their own assets primarily for counterfire missions, and one BCT developed a mature enough set of fire support coordination measures to utilize a Free Fire Area for both artillery and attack aviation in the insurgent-controlled palm grove areas.[275] As the counterfire effort in Baghdad matured, fire supporters at higher echelons began to integrate and influence operations. This attempt to integrate and streamline target acquisition and the clearance of fires led to many perceptions of micromanagement at lower levels. Although one counterfire effort in 2007 reduced rocket attacks from a peak of 20 per day at one location in eastern Baghdad, the commander expressed his frustration at not being able to control his own 120mm mortars that provided the counterfire.[276]

The ensuing permissive environment for indirect fires and years of institutional learning combined to produce the most refined employment of artillery units in OIF. In the Baghdad belt areas in 2007 to 2008, commanders used a small element of the artillery battalion to provide indirect fires while the rest of it conducted the same security mission set as other infantry units, albeit on a smaller scale.[277] Using artillery units in a split fashion was nothing new, as Odierno's artillery battalions in Salah-ad-Din all simultaneously provided fires and secured an area as a maneuver task force in 2003.[278] One BCT southeast of Baghdad used a small detachment of "a couple of pieces" available to deny selected terrain and fire on targets of opportunity, using split sections to increase coverage and responsiveness across their area.[279] Another BCT to the southwest of Baghdad maintained a four-gun firing element at one of their FOBs, using it for an active counterfire campaign of both lethal fires and illumination to disrupt enemy operations.[280] The next unit in that area of operations kept a battery at that location, and attached both an infantry company and a mechanized company team to act as a maneuver task force for security.[281] These situations called for adaptive leaders and flexible command relationships. In one case, a battery commander reported to one maneuver commander for issues in his area of operations as a maneuver task force, his original commander from the artillery battalion for issues with the platoon providing the BCT's indirect fires, and an additional support structure from another BCT to support his logistics.[282] Additionally, artillery units with mixed roles in the Baghdad belts were able to dynamically task organize for specific circumstances in their area. One battery commander took the minimum amount of personnel out of his regular patrol cycle so they could establish firing capability, which effectively disrupted IED attacks along a critical stretch of highway for two weeks.[283]

The employment of artillery units in Basra was on a similar arc to that of coalition forces in the rest of the country. After the initial ground invasion, British forces sent their guns home and gunners spent the next two years patrolling to augment infantry units due to the relatively small size of the British contingent compared to other coalition forces.[284] Artillery units were continually employed in security roles and to train ISF soldiers until indirect fires attacks against the operating base in Basra required the establishment of a counterfire capability. Some British officers viewed this shift as a cognitive disconnect, echoing the notion at the time that the British presence itself was engendering violence instead of the large JAM influence in the city. One British artillery officer commented that "[w]e were trying to hand over the four provinces, but we were trying strike operations against the IDF teams. We'd all withdrawn to the COB, we were the problem. If the COB hadn't been there, there'd have been no attacks."[285]

This led to a situation which greatly constrained the use of artillery fires to protect the force. Since any violence after *The Accommodation* during the withdrawal of British troops from Basra city had large political implications, authority for counterfire missions was held at the highest level possible, the national command authority.[286] Although there was a large requirement for counterfire in the rural areas of MND-SE, the howitzers remained near Basra where they were relatively hamstrung by these political limits. When Operation CHARGE OF THE KNIGHTS began, British forces regained the same conditions-based counterfire capabilities that American forces learned to use in Baghdad in 2007. Fire supporters and targeteers proactively estimated collateral damage concerns and developed an overlay of Basra. Commanders were able to leverage air assets that MNF-I sent to support Operation CHARGE OF THE KNIGHTS in areas that required more precise targeting.[287]

By the security transition in 2009, coalition forces' employment of artillery fires consisted of small-scale support to ISF units, particularly Federal Police units in urban areas. ISF units were able to leverage mortar fires from AABs, but only for deliberate missions.[288] Fires battalions within the AABs provided another battalion-level headquarters to partner with ISF.[289] General support artillery units at the enduring FOBs outside of urban areas were generally the only source for indirect fires. They found use as a means of demonstrating that ISF still had the backing of American firepower.[290]

Throughout OIF, commanders employed artillery units in a nonstandard role of security, often operating as a maneuver task force with its own area of operations. Following the ground invasion, some artillery units were immediately pressed into service in security roles since there was about a nine-month lull until artillery fires were widely used again.[291] As the war continued and commanders increasingly sought to enhance security in their area, artillery units were a logical choice due to their identity as a combat arms element. Some units came to rely on them as a maneuver force to the point that they only conducted minimal traditional artillery training between deployments to OIF.[292] One battalion commander noted that artillery units are an appealing option since they already have the basic qualities of a maneuver task force such as small arms proficiency, combat vehicles, and secure communications.[293] Critically, artillery units contain trained combat arms officers with the ability to integrate effects, with an exposure to the fundamentals of fire and maneuver through their experiences as fire support officers. Battery commanders serving in the maneuver role combined these qualities to a great effect in many areas.[294] However, manning and equipping these forces was a key issue since an artillery unit is much smaller than an infantry counterpart, with a comparatively lean staff and support element.

During the counterinsurgency effort, artillery units serving in a maneuver role learned to harness their core competencies positively. One BCT commander remarked that he viewed artillerymen as particularly good at counterinsurgency because they are usually tasked as a supporting role to the main effort in conventional operations. As such, they have no hesitation or reluctance when supporting an ISF or local leader instead of seeking to act as the main effort in that area.[295] Another battalion commander, who successfully converted his artillery battalion to conduct a nonstandard security mission, noted that artillery units have a unique approach to counterinsurgency operations:

> I think that artillery units are successful in COIN because they don't carry any baggage from having conducted operations at one end of the spectrum and are then asked to transition to a COIN mindset. We don't have a specific mindset to begin with. It is much easier to teach a unit how to take on a new role because there are no bad habits to un-learn. In that sense we were blessed.[296]

Many commanders also tasked artillery units with the complicated mission of training ISF units and advising combat operations before the MiTT model was adopted across Iraq. MNSTC-I deliberately delayed raising and training IA artillery capabilities. This became apparent in 2007 when Petraeus sought to accelerate national programs to give ISF indirect fires capabilities, but MNSTC-I did not have an artillery liaison unit to even plan this contingency.[297] Without Iraqi artillerymen to train, artillery units were able to teach marksmanship, basic soldiering skills, and small unit tactics to the ISF by virtue of their combat arms training. 1st BCT, 1st Armored Brigade used their artillery battalion to operate an ISF training center as part of the effort to pacify Ramadi, which was key due to the massive influx of new policeman from the tribes.[298] In this model, policemen would train at distant centers in Baghdad or Jordan to learn core competencies, then polish their combat skills at Camp Phoenix in Ramadi since the commanders there recognized that in a counterinsurgency "police needed to be able to fight like infantry squads due to the kinetic nature of the fight."[299] The key component of that arrangement was that artillerymen were used to train a skill set that they were familiar with themselves. On occasion, artillery units would be employed in a stopgap training support mission, as in the case of one battalion that was detailed to train policemen on their actual policing tasks:

> If our solution to policing is to re-designate or dual-purpose artillery units to do that, then we are not holding ourselves accountable. We are coming up with a Band-Aid solution to something that is far more complex. In my 30 years as an artilleryman, there wasn't a lot that would have helped me develop police in a culture that I was fundamentally ignorant of.[300]

This quote reveals an important consideration in the employment of artillery units in nonstandard missions. If specialized troops for a unique or unforecasted mission in counterinsurgency do not exist, the army must make a parallel effort to resource this while an *ad hoc* organization or re-missioned unit maintains steady improvement.

Commanders only embraced this concept marginally in OIF, and many artillery units found themselves in unforecasted, yet enduring, mission sets which required emergency training and adaptive leadership.[301] Artillery units performed diverse tasks such as civil affairs and reconstruction support, convoy security, processing Captured Enemy Ammunition (CEA), and company-level intelligence support. In 2003, much of the initial focus on reconstruction efforts meant that the undermanned civil affairs operations required support. 1st Armored Division used their general support MLRS battalion to control the security and reconstruction efforts at Baghdad International Airport.[302] Some BCTs assigned their fire support elements to assist civil affairs unit's local-level efforts in an arrangement often referred to as "Team Village."[303]

Other artillery units conducted operations to locate, process, and backhaul CEA. Due to an artillery unit's built-in capacity to haul ammunition and knowledge of considerations for transport and storage, they are particularly well suited for this mission set. In 2003 this was extremely important since the old Iraqi army left huge depots of large-caliber ammunition which could be easily exploited by insurgents to produce IEDs.[304] Since units were finding everything from World War II-era ammunition to French Roland missiles, one Division Artillery commander whose units were processing CEA said it was more like "demilitarizing" a society rather than reconstruction support.[305] A nonstandard mission of convoy security was similar to the easy transition to CEA operations, since it built on a core competency of artillery units. Since artillery units secure themselves during most mounted movements on a conventional battlefield, it was not a stretch to use them to secure another element such as a long-distance logistics convoy or a construction project site visit. Improvements in logistics units' ability to secure themselves removed much of the burden for this, and was reduced further in the use of contractors to secure elements of the Department of State and the Corps of Engineers. However, the need was so great in Baghdad in October 2007 that one artillery battalion was re-tasked to secure embassy personnel at the temporary cost of indirect fire support to an entire BCT.[306] Although artillery units were not specifically tasked to man company-level intelligence support teams, many commanders used their assigned fire support personnel to establish this cell. By 2007 this was widely accepted as a best practice, due in part to a fire supporter's familiarity with targeting procedures and combat operations with the maneuver force.[307]

Analysis

In broad terms, the counterinsurgency effort in OIF suffered from a disconnection between the markers of success at various levels of the conflict. Counterinsurgency is a junior commander's war, meaning that the effort to defeat an insurgency can only be decided among the population at the local level. There is no strategic silver bullet, no institutional knockout punch. However, a cogent operational approach to pursue the strategic objective must be present to make these victories on the street count toward a steady cumulative effect. Chiarelli captured this in a 2005 writing, reflecting that "[t]he broad collection of small, decisive victories along all lines of operations, supporting each other in a delicate balance of perception and purpose, would move the campaign toward positive results."[308] The disunity of efforts between MNF-I and the GoI, and within MNF-I in OIF were a symptom of this phenomena, not necessarily a cause.

The lack of a unified effort between the coalition and the GoI was evident early in the counterinsurgency effort, almost as soon as the central government took shape. Operations such as the efforts to clear Fallujah in 2004 were important signals of the coalition's resolve, but they also disrupted political reforms since they provided Sunni insurgent groups with a convenient excuse for political retrenchment.[309] Early efforts to partner with elements of the GoI were focused almost entirely on security sector reform; only with rare exceptions did American units partner with the other critical links of Iraqi governance and civil stability such as resource distribution networks.[310] Many early efforts to harmonize the counterinsurgency effort between MNF-I and the GoI failed because much of the joint campaign plan for rebuilding Iraq was classified, relied on third-country contractors, and did not involve Iraqi leaders in the planning process.[311] Security efforts suffered similar issues, and senior ISF commanders' fundamentally different read on problem sets were under-informed or ignored until Odierno provided significant clarity to operations in 2007.[312] This was solidified by the psychological effect of *The Surge*, which displayed sufficient resolve to support the GoI.[313]

MNF-I's counterinsurgency effort also suffered from the lack of an internal unified effort between military and diplomatic elements, between conventional and special operations units, and within the effort to develop the ISF. The most significant of these was the disarray between military and diplomatic elements, since the crucial task of rebuilding the institutions of governance in Iraq institutionally fell between them. From the outset of the counterinsurgency effort, Garner admitted that ORHA could not integrate these functions without complete security for his civilian elements. This led to a huge unforecasted bill for military security.[314] Tensions between the military continued as ORHA transformed to the CPA, to the point that Bremer was told by military officers that he'd "stepped out of his lane" when he questioned their ability to find effective means to combat the enemy.[315] During this organizational divide, military Civil Affairs units had a limited effect, and in many areas the integration of these efforts was only a priority when the local commander saw it as such. The heavy use of security contractors by CPA and the Department of State clearly exacerbated issues. Many unit commanders derided the conduct of these contractors since they were strictly focused on securing their principal, and as such they had no stake in avoiding collateral damage or respecting the local community. But critically, most Iraqis did not understand this context and viewed these contractors as agents of the local American military unit.[316]

Steady improvement in coordination began when the PRT concept transferred from Afghanistan in late 2005. When the Department of State began to fully resource these teams and they embedded with specific units in their area of operations, it served to unify the effort down to the BCT-level in many areas. Embedded PRT leaders often had a seat at the table with their BCT commander counterparts, and were able to integrate their core competencies into staff processes.[317] While the PRT model had success in physical reconstruction efforts locally, some advisers serving in headquarters at higher echelons did not see the same effect on efforts to rebuild viable nationwide government institutions.[318]

The military component of the counterinsurgency effort also experienced internal discord among many conventional and special operations units. Interestingly, this is perhaps the closest an American-led coalition has ever come to Trinquier's *quadrillage* model of counterinsurgent task organization, with conventional units (both coalition and ISF) serving as the grid troops, Special Forces teams as interval units, and special operations elements functioning as intervention units to focus on insurgent leadership. However, there is not ample evidence that this was ever a deliberate framework of operations, and as a result there was tension in some areas that conventional unit leaders saw as their own when special operations forces operated outside of their control or influence.[319] Special Forces leaders experienced similar tension, and resented the fact that some conventional leaders thought that they were only focused on targeting the enemy.[320] Quite precociously, one Special Forces commander posited that the key was managing the varied personalities in OIF: "relationships between commanders are more important that command relationships."[321]

The lack of a unified effort to develop and advise ISF units also created issues at the local level. At a certain point in a counterinsurgency, there may be a threshold point where the organization of combat units is revised to support an advisory mission set instead of a security mission set. Partnering units is a generally effective means of improving security forces since commanders at all levels can empathize and develop rapport over time. In OIF, a MiTT was a valuable addition to mentor and advise at a higher command echelon, and also provided a communication channel to MNSTC-I's national-level effort of raising and equipping ISF elements. However, some coalition commanders cited the danger of advisory teams grading their own ISF unit's progress.[322] The AAB concept sought to blend the best qualities of these elements under one cohesive tactical command, and in large part it received favorable reviews for that task. By taking advantage of the inherent efficiencies of existing leadership and robust staffs within a BCT, the AAB model was effective. But MNSTC-I could have pushed to introduce it in regions outside of Baghdad much earlier.[323]

If a unity of effort describes the harmonization of the counterinsurgency laterally across many functions, the concept of "nesting" describes the harmonization of efforts vertically to ensure that successive echelons support a higher objective. To this end, several control measures that were traditionally used on a linear conventional battlefield were cognitively modified by adaptive leaders to fit the demands of a volatile and amorphous environment in Iraq. Issues with traditional unit boundaries based in geographic terms caused many unnecessary issues, especially in the Baghdad belts where some units would have to nest their concepts and purpose with the MND-B commander but report to another unfamiliar division-level commander.[324] Efforts to synchronize effects between the BCT and battalion levels were generally easier to accomplish since far more units in this linkage had an organic relationship from home-station training. One BCT operations officer described this as giving battalions minor "course corrections" to ensure their well-informed efforts at a local level were tied to the BCT commander's vision and objectives.[325] This synchronization helps this effort, but it entails more than just assigning subordinate commanders an area of operations.[326] Field grade officers can form the supporting cast in a counterinsurgency, and manage support functions to include intelligence, command and control, communications, and logistics to support a commander's understanding and decision-making.[327] The Taji COIN Center provided another measure of harmonizing operations, since all incoming units in 2007 and after studied commanders' intents at multiple levels of the counterinsurgency effort. One commander said that since he understood these intents, he could adapt methods to a changing environment in order to obtain that endstate within his area.[328]

Throughout OIF, the counterinsurgency effort struggled to defeat insurgent groups within a framework of legitimacy. This will always be difficult for an intervening country, which was the case in OIF as soon as the GoI was granted sovereignty in June 2004. The counterinsurgency effort had to overcome additional difficulties since many groups such as Kurds and Sunni Arabs did not necessarily view a Shia-dominated GoI as a legitimate power over them. For MNF-I, this created an environment of balancing cooperation with anti-GoI elements for short-term security gains with efforts to create a social linkage via reconciliation between disaffected groups in the long term. Besides political legitimacy, which was found in partnership and eventual subordination to an effective GoI, there was also an enduring issue of legitimacy in the eyes of the people. After the first year, coalition forces had difficulty countering the narrative that they were foreign occupiers. They could only focus attention on the fact that an occupation does not entail efforts to improve the population's lot in life, and draw attention to those successful coalition actions which accomplished this end.

In attempts to isolate insurgents from their base of support, the counterinsurgency effort adapted to the environment in OIF and in effect isolated the base of support from the insurgent. Across most areas in Iraq early in the war, large sequential cordon-and-search operations took the place of large sequential search-and-destroy missions from Vietnam. These tactics were similar in the way that they generally failed to hold or build key areas and population centers, consequently failing to separate insurgents and their base of support. In general, since insurgents did not mass except immediately prior to an attack, they did not have consolidated resource requirements. Because a small cell within a network only required a few caches and a few bed-down locations in a large urban area, massive sweeps without extremely detailed intelligence did not disrupt them.

Compartmentalizing populations with permanent cordons and operating within those cordons proved effective in achieving the ultimate effect of separating insurgents from the one resource they inherently lacked: targets. In the Baghdad of 2006, both AQI and JAM did not necessarily seek to defeat ISF nodes, although their neutralization and an ensuing lack of security was advantageous.[329] What both groups truly required from the population was Shia mosques and markets for VBIEDs or Sunni neighborhoods for intimidation and killings. Truly isolating insurgents from their base of support came from security gains once trusted local political powers and local irregular security forces could assert control within an area. Population security is really the distillation of this technique to its basic element.[330]

Operations to clear insurgents with armed force are closely related to this technique. Operations in Baghdad in 2007 highlight that clearing insurgents with armed force does not stop at killing the enemy; it must rescue the population.[331] In OIF, the counterinsurgency effort showed the misguided use of this principle in the early years of the war since killing the enemy took precedence over the follow-on task of securing the population, and sometimes completely obviated it. Colonel Craig Collier, a former battalion commander, asserted that "our experience in Iraq verified that lethal operations remain the decisive element of combat power."[332] This assertion reveals some faulty underlying assumptions which led to the simplistic effort to clear insurgents solely with lethal combat power. While lethal operations have a definite place in a counterinsurgency, combat power is not decisive in and of itself in that form of warfare. Insurgents will only remain cleared if the population sees it as a distributive outcome, one with a permanently viable solution. Collier continued, claiming that "[w]e seem reluctant to admit that killing the enemy worked."[333] Killing the enemy always works in the short term. Although there could be many second-order effects in a hypothetical situation, killing the enemy generally makes for one less insurgent if he is killed judiciously. But simply killing the enemy is not enough, and as successful programs in Iraq demonstrated, it is much better to co-opt an enemy because a dead enemy cannot give you additional security, HUMINT, or spend his wages locally. This notion should not dissuade well-informed, pragmatic, and episodic efforts to clear insurgent bases with armed force. As the second operation to clear Fallujah illustrated, the act itself clearly signals resolve in the counterinsurgency effort, and can make insurgent elements wonder if they can really outlast it.[334]

Within the context of these challenges, *The Surge* (and OIF to a larger extent) was not a revolutionary approach or a watershed moment of institutional enlightenment for the counterinsurgency effort. Instead, it was the confluence of several important factors that restored synchronization, continuity, and a unity of effort. Crucial additional forces arrived when two adaptive tactics were implemented in Baghdad and the belts: a method to co-opt tribal local power structures and integrate them into the security framework as in Al Qaim and Ramadi; and a method to secure the population during selective strikes at insurgent groups as in Tal Afar. These tactics were not derived from a theoretical framework, they were the outgrowth of four hard-fought years of iterative experience in counterinsurgency warfare by American and British forces. Odierno's operational approach and Petraeus' theater-level strategic framework enabled the restoration of a unified counterinsurgency effort, but it would not have succeeded without these refined tactics and sufficient troops, resources, and political backing from the institutional military.

A complete analysis to describe the factors that influenced the employment of artillery units during OIF are confounded by an important variable. OIF was concurrent with the Army's shift to the modular BCT model, in which artillery battalions transferred from the Division Artillery to specific BCTs.[335] As a result of this transformation, artillery units maintained their branch-specific proficiency and standardization through informal relationships and division-level initiatives, but struggled to replicate the functions of senior artillery mentorship and external evaluations.[336] What distinguished this process was not so much the ruminations of an Old Guard of senior artillery officers, but organizational resistance to change within the junior officer population.[337] Some legacy brigades completed the transformation to BCTs as late as 2006. This introduced additional organizational friction and uncertainty over the nature of command relationships and options to employ artillery units. Quite simply, BCT commanders did not have much experience commanding them.[338]

The requirements for tactical fires were a factor in the employment of artillery units, but a very localized one. Insurgents rarely massed to provide lucrative targets, except in pre-planned complex attacks. Indirect fires were used on these rare occasions, but not generally in sustained operations except for the large-scale clearances in Fallujah or Diyala. Opportunities to support troops in contact were relatively rare, since insurgents conducted ambush attacks and sought to disengage immediately. A typical engagement lasted under a minute.[339] As with Malaya, most insurgents sought to avoid contact before and after an ambush to negate their disadvantage in numbers and firepower. Deliberate targeting did not provide much of a stimulus for indirect fires beyond the aforementioned H&I fires to deny terrain. In many cases, focused raids to kill or capture insurgents replaced the intelligence-based fire mission. Planning for raids took on much of the same character as targeting for a fire mission with functions for corroborating target intelligence, target selection standards, and attack guidance. Since HUMINT was at such a premium in OIF, selective strikes against insurgent locations would not have been sustainable since intelligence gleaned from sensitive site exploitation in one raid could enable many more operations.

The effect of organizational constraints and physical limitations in OIF had a greater effect on the employment of artillery units. Constraints reflected the effort to avoid collateral damage, which some officers viewed as creating more insurgents.[340] It was a fine line, which Captain Jim Kiersey, a company commander in 2007, summarized by saying that "[s]hooting the right guy teaches the enemy and population that evil has consequences; the corollary is that a poor shot – one that hits an innocent person or leads to collateral damage – is worse than not shooting at all."[341] To minimize collateral damage, units implemented the practice of Collateral Damage Estimation (CDE) to inform these decisions. Although it was an additional step which some considered an annoyance, it was effective and rarely led to a denied mission request.[342] Not all commanders understood this distinction, leading one British officer to contend that "I'm not quite sure if the maneuver units understand the low CDE you get with GMLRS or other low-yield weapons. It's an imperception among us that if you use artillery in COIN, then something is going wrong."[343] As a result, many measures were delegated to tactical commanders to alleviate the restrictive or unresponsive nature of CDE validation.[344]

The physical limits in Iraq were practically negligible when compared to Malaya and Vietnam, but there were some factors which influenced the employment of artillery units. The terrain itself offered few limitations, and the issue of accounting for meteorological conditions at remote locations was mitigated by a technique from operations in Afghanistan to convert readily available Air Force data.[345] While the urban terrain in many areas of operation limited the direct observation of targets, many units were able to utilize a vast array of air and ground sensors to track targets. Since Global Positioning Satellite data was available to all patrols and many of them had advanced optics, target location error was greatly reduced from previous counterinsurgencies.

The counterinsurgency effort was task-organized in such a manner that there was a large demand for additional security forces, and this was one of the primary factors influencing the employment of artillery units. In the words of one commander, in a counterinsurgency "you never have enough forces."[346] Until significant numbers of ISF units were reliable enough to take responsibility in some areas, the counterinsurgency effort was stretched so thin that some planners viewed almost everywhere outside of Baghdad as an "economy of force effort" until 2008.[347]

Although there was a relatively dependable amount of air support for emergency situations, it was never a strong enough force to push artillery units exclusively into nonstandard security roles. Commanders were confident in Close Air Support, but only for crisis situations with troops in contact against an insurgent element.[348] Attack aviation assets were also dependable in OIF, and in some cases were used for counterfire missions when the use of artillery fires had been denied by a higher headquarters.[349] Helicopter assets provided an excellent mix of additional surveillance, reconnaissance, and firepower for a ground commander but they were limited by force protection concerns around urban areas and loiter time away from refuel points at the larger FOBs. Some commanders saw an overreliance on air assets to be an unnecessary risk when artillery units were available.[350] Air assets were never able to fully replicate the dependability of artillery fires, and when one BCT secured a new area of operations the commander employed part of his artillery unit for indirect fires because "[w]e were going somewhere nobody had been, so you'd better be able to take care of yourself."[351]

The repeated cost of converting artillery units between indirect fires and nonstandard security forces was not only a major factor in their employment during OIF, but also a long-term issue for the artillery branch of the American military. The combined requirements of artillery units in OIF and combat operations in Afghanistan created some artillery leaders with the experience of firing thousands of fire missions in counterinsurgency combat, some artillery leaders with more experience as a maneuver commander than their infantry branch counterparts, and a majority of artillery leaders with a mix of both experiences. There is a sense that these repeated conversions in and out of indirect fire roles created a branch that is the metaphorical jack-of-all-trades but master of none. In many cases, the critical unit-level lessons learned have not been collected, let alone exploited. This created a large deficit in the institutional base of knowledge within the artillery community, a deficit that will take many years to reduce.

Three veteran BCT commanders recognized this trend in 2008 and wrote a five-page white paper titled *The King and I: The Impending Crisis in the Field Artillery's Ability to Provide Fire Support to Maneuver Commanders* which stated that they had "watched the deterioration of the Field Artillery branch with growing alarm."[352] Their well-sourced opinion paper recognized the cumulative effect of nonstandard artillery unit employment and the loss of senior artillery mentors in the new modular BCT construct. Their evidence stemmed mostly from substandard performances of artillery units in their traditional role at the combat training centers.[353] The influences of training and educating leaders, training units, and converting their internal structures for nonstandard roles contributed greatly to this state, and continued to be a factor in their employment in OIF.

The first challenge in converting an artillery unit to a maneuver force or another nonstandard security role lies within the unit's leaders. Due to the artillery's intense requirements for technical and tactical competency, some commanders felt that their artillery units' leaders suffered from a "checklist mentality" initially, when they needed to adapt to different contingencies in counterinsurgency.[354] Between focused training and exposure to missions among the population as fire support officers, artillery units were able to prepare their leaders to maintain an adequate result for the initial 90 days of a deployment and adapt from there.[355] The following quotes illustrate two commanders' reflections on the method, and the importance of training artillery leaders for the complex and lethal counterinsurgency environment of OIF:

> We were turning artillery platoons into infantry platoons, artillery sections into infantry squads, so it is a little bit of a round peg into a square hole. But, where we could, we kept the NCOs and officers in place so that there wasn't the added stress of having to learn a new organization.[356]

> You have to bring the right leaders in. You have to look for ways; it all goes back to leadership and you have to know your organization, the psychology of combat, and then apply your understanding of the organization to that environment.[357]

The effort to convert artillery leaders for nonstandard roles in the counterinsurgency effort was largely successful.[358] However, the effect of repeated out-of-role deployments took a toll on many artillery leaders. Some officers reported that their artillery counterparts were actually much more comfortable training as maneuver forces after their deployment, since it was what they knew and understood.[359] One BCT commander realized that his company-level fire supporters knew more about using money as a weapon system than artillery as a weapon system since they were leading company intelligence support teams instead of de-conflicting airspace and echeloning fires.[360] The effect on junior NCOs was also evident, as some were reaching the point of centralized selection boards without ever working a fire mission on the gun line.[361] One artillery battalion commander was faced with the challenge of retraining his unit after it operated for nearly five years as a maneuver task force. He recognized that the first step in regaining capable leaders was competency (gained through technical certifications), and then with confidence (gained through repeated training and evaluations).[362] A battery commander with a similar situation recalled that "it reinvented the wheel for guys who had been away from firing operations for four to five years. It was a focus of individual section skills, section certifications, and a platoon EXEVAL with the Fires Brigade."[363]

Training artillery units in their new roles was another challenge. This was exacerbated by the difficulty in replicating a complex counterinsurgency environment in Iraq, since combat training centers could only replicate certain facets.[364] A key to adequately preparing platoons and batteries for their next mission was predictability. Units that knew their mission for the next deployment could develop detailed training plans and begin to harness their internal lessons from previous operations. Even one AAB that decided to keep a platoon trained for indirect fires was able to make that decision with enough lead time to allow the platoon time to conduct all live firing tables prior to their mission rehearsal exercises.[365] With predictability, one benefit of converting an artillery unit to a provisional security element was that they were purposely tailored for a specific mission profile, based on a unique area of operations and its peculiarities.[366]

When training artillery units to convert back to delivering indirect fires, proactive units were able to integrate artillery training during continuous operations in Iraq. Not only did this maintain a minimum level of preparedness in case the tactical situation called for a return to firing artillery, it also significantly shortened the time it took to re-train and certify upon redeployment. Periodic training on the most basic skills such as maintenance checks on idle howitzers mitigated the decay of those skills.[367] Units were even able to reestablish firing capability during deployments without sacrificing the volume of patrols in their area.[368] Upon returning home, units struggled to regain firing capabilities if most of the junior leaders changed units or immediately left for professional schooling that was delayed for the long deployments. Losing the experienced battery commanders and platoon leaders left a void for planning and resourcing training plans, and losing experienced NCOs left a void of resident technical knowledge and certifications.[369] Units were forced to use all available artillery experts, to include external observers and mobile training teams. One battalion relied on its commander to certify firing units prior to their deployment to Afghanistan in a firing role:

> At the end of the day, we'd leave them and we'd actually head up to the OP to watch, hope and pray that it made it in the impact area. It was great because it instilled confidence in them, the coaching piece was huge. I can say I was able to take 18 years of artillery stuff and impart it onto LTs and E-6s. I was confident that they'd be able to provide fires and not have a firing incident.[370]

The third effort to convert artillery units for nonstandard roles was the structural conversion of staffs and equipment distributions. This was an important nuance in the employment of artillery units, since headquarters at higher levels could confuse a unit's relative combat power based on the echelon of their representative icon. Without significant augmentation, a battalion icon for an artillery-based maneuver task force was not equal to an infantry task force with the same mission set.[371] A key area for augmentation was on staffs, specifically the intelligence section. Most artillery units did not have a robust intelligence section since artillery intelligence traditionally does not use HUMINT as an input. This was a particularly glaring weakness in a counterinsurgency campaign where quality HUMINT is one of the most precious commodities.[372] Similar to the ability of quality leaders to adapt to a different mission, quality staffs were also able to adapt with just a few key augmentees. One artillery battalion commander in OIF remarked that:

> I think because we had a well-trained artillery battalion staff and we were proficient at the targeting process, we were able to adapt our staff functions to be able to run battalion operations in a COIN environment without a heck of a lot of additional effort.[373]

Successful leaders also matched existing core competencies with required capabilities, such as a fire direction center's ability to process and analyze data lending itself to a battery intelligence support team.[374] Since artillery units required augmentation, they maintained most of their own equipment and personnel. This meant that they had relatively fewer challenges to structurally convert back to delivering indirect fires. The only hindrance was the lengthy reset of howitzer equipment after a deployment, but many commanders were able to temporarily transfer equipment to enable training since "there's no substitute for going out and firing to regain gunnery skills."[375]

The effort to maintain indirect fires proficiency throughout OIF became a concern for some senior leaders to include Petraeus, who recognized that the artillery branch's commitment to high standards of responsiveness and accuracy made the enterprise demanding even before the stresses of OIF.[376] One of the shortcomings in this endeavor was the failure to promulgate lessons learned from OIF, or at least avail them through accessibility.[377] Without robust networks to promulgate the lessons of all roles in counterinsurgency, leaders relied on periodic professional journals and cross-leveled experiences as individuals changed units immediately after redeployment.

Conclusion

In the introduction to his narrative of operations in Baghdad during 2007, Kilcullen summarizes the incredible challenge to the counterinsurgency effort as follows:

> If we were to draw historical analogies, we might say that operations in Iraq are like trying to defeat the Viet Cong (insurgency), while simultaneously rebuilding Germany (nation-building following war and dictatorship), keeping peace in the Balkans (communal and sectarian conflict), and defeating the IRA (domestic terrorism). These all have to be done at the same time, in the same place, and changes in one part of the problem significantly affect others.[378]

Although the coalition forces formally transferred security and governance responsibilities to a viable host nation apparatus, it is erroneous to characterize OIF simply as a successful counterinsurgency. A more inclusive characterization is that OIF was a war in which an adaptive and resolute effort by both Iraqis and coalition forces achieved a successful endstate.

The role of artillery units in this war illuminates two key points regarding the prosecution of a counterinsurgency effort. The success of artillery units serving as maneuver task forces shows that counterinsurgency is not an endeavor of pure combat might, where only trained infantry or cavalry elements can execute an enemy-focused operation. Counterinsurgency requires a highly nuanced approach to violence which the artillery branch can balance, despite its traditional penchant for overwhelming firepower and large explosions. The concurrent success of artillery units serving as reliable providers of indirect firepower shows that in counterinsurgency, violent force is still a capability that must be used to balance a successful approach.

Notes

1. This chapter's epigraph is from T. E. Lawrence, "The 27 Articles of T.E. Lawrence," *The Arab Bulletin* (20 August 1917), Section 15. Recommended secondary sources for further study in OIF include (in rough chronological order of their themes): Michael Gordon and Bernard Trainor's *Cobra II*, Thomas Ricks' *Fiasco* and *The Gamble*, and David Kilcullen's chapter "The Twenty-First Day," in *The Accidental Guerrilla*. For a concise overview and analysis of the entire war, refer to Carter Malkasian's "Counterinsurgency in Iraq: May 2003-January 2010," in *Counterinsurgency in Modern Warfare*. For studies regarding the successes and failures to innovate and adapt during OIF, see Daniel Marston's "Adaptation in the Field: The British Army's Difficult Campaign in Iraq," in *Security Challenges* (Autumn 2010), and James Russell's *Innovation, Transformation, and War: Counterinsurgency Operations in Anbar and Ninewa Provinces, Iraq, 2005-2007*. For the foreseeable future, there will be a lack of credible, unclassified documents from the insurgents' point of view, consequently Ahmed Hashim's *Insurgency and Counterinsurgency in Iraq* is the most thorough resource for understanding the roots of the insurgency, and specific insurgent groups.

2. Brian Burton and John Nagl, "Learning as We Go: the US Army Adapts to COIN in Iraq, July 2004-December 2006," *Small Wars and Insurgencies* 19, No. 3 (September 2008): 323.

3. David Kilcullen, *The Accidental Guerrilla* (Oxford, UK: Oxford University Press, 2009), 148-151.

4. Kilcullen, *The Accidental Guerrilla*, 117.

5. 1st Infantry Division, *Soldier's Handbook to Iraq* (Wurzburg, GE: 1st Infantry Division, 2004), v. The 1st Infantry Division issued this handbook to soldiers before deployments to OIF in 2004, and it is representative of handbooks developed by most US Army divisions or brigades in the earlier years of the war. While these handbooks are thick with background facts of Iraq and useful everyday phrases, they are relatively thin on contemporary analysis of culture in Iraq.

6. 1st Infantry Division, v. All demographic data reflects estimates of Iraqi society in 2003. Unless otherwise noted the term *Sunni* will refer specifically to Arabs in this chapter, even though most Kurds and foreign fighters were Sunni as well.

7. BF020, Civilian Adviser to MNF-I, Interview by Richard Johnson and Aaron Kaufman, Boston, MA, 11 March 2011. This notion must be tempered by the fact that, due in large part to Saddam Hussein's brutal regime, both the Kurds and Shia Arabs did not understand how American soldiers of different ethnic backgrounds could fight together for a common goal. Although these groups did not completely advocate communal violence at first, they did not necessarily understand cooperation. See also: Michael R. Gordon and Bernard Trainor, *Cobra II: The Inside Story of the Invasion and Occupation of Iraq* (New York: Pantheon Books, 2006), 441.

8. 1st Infantry Division, v. These ancient urban centers affected the very formation of the country. Modern Iraq was formed from the area common to the

Ottoman Empire's provinces of Mosul, Baghdad, and Basra, which were based on the three major cities and a surrounding Governate rather than distinct areas of tribal, ethnic, or sectarian populations. Charles Tripp, *A History of Iraq* (New York: Cambridge University Press, 2005), 8-9.

9. BF020 Civilian Adviser to MNF-I, Interview.

10. James Aylmer Haldane, *The Insurrection in Mesopotamia* (London: The Imperial War Museum, 2005), 39. As the General Officer Commanding, Haldane's account of operations in early Iraq illustrates the difficulty of military action among the tribal structure. As the British were still deliberating whether a system of direct or indirect rule would be the most effective system in Mesopotamia, a revolt began almost immediately in 1920. As British authority was restored, the administration recruited for government offices and the fledgling army officer corps almost exclusively from the Sunni Arabs, even though they comprised less than 20 percent of the population at the time. Tripp, 30-31.

11. Charles Tripp, *A History of Iraq* (New York: Cambridge University Press, 2005), 77.

12. Tripp, *A History of Iraq*, 143. Although the Ba'ath Party became a vehicle for Sunni elites to consolidate power, it was actually founded by a Shia in An Najaf.

13. Tripp, *A History of Iraq*, 190-193. Throughout the 1970s, this regime consolidated its domestic power and grew to a relative position of strength in the Middle East by leveraging its vast oil resources against the volatile world economy.

14. Tripp, *A History of Iraq*, 233, 238-239. The Shatt al-Arab is the large waterway formed by the confluence of the Tigris and Euphrates rivers, providing Iraq with its only seaports and maritime access. These attacks were in direct violation of a 1975 treaty between Iran and Iraq, which clarified the southern and western shores for Iraq and the northern and eastern shores for Iran. This marked the first instance of Hussein's unprovoked attacks which would be cited as a political pretext for pre-emptive invasion in 2003. A large and successful Iranian counterattack in 1982 led to a long war of attrition through 1988. French assistance to develop an air force and American naval presence in the Gulf persuaded Iran to settle for peace, and the United Nations (UN) monitored the eventual cease-fire.

15. Bernard Lewis, *The Crisis of Islam: Holy War and Unholy Terror* (New York: Random House, 2003), xxiii.

16. Tripp, *A History of Iraq*, 251.

17. Tripp, *A History of Iraq*, 252-254. Although he used a dubious historical claim to the annexation of Kuwait, this was simply a narrative for the international community's consumption.

18. Thomas E. Ricks, *Fiasco* (London: Penguin Press, 2007), 5-6.

19. BF010, Former Army Officer, Interview by Richard Johnson and Aaron Kaufman, Boston, MA, 11 March 2011.

20. Ricks, *Fiasco*, 13-15. This is illustrated by the fact that the US Air Force did not lose a single aircraft in 12 years of enforcing the no-fly zones.

21. Ricks, *Fiasco*, 32-33. CENTCOM began to draft war plans as early as November 2001, and the CENTCOM Commander, General Tommy Franks, briefed President George W. Bush in December 2001. For Franks' account of this meeting with the President, see Tommy Franks, *American Soldier* (New York: Regan Books, 2004), 346-356.

22. Ricks, *Fiasco*, 49. This led Vice President Dick Cheney to publicly assert "there is no doubt" that Iraq had WMDs during the run-up to war. For additional analysis on the doctrine of pre-emptive military actions, see: Andrew Bacevich, *The New American Militarism: How Americans are Seduced by War* (Oxford, UK: Oxford University Press, 2005), 147.

23. BF010, Former Army Officer, Interview. Additionally, critics asserted that the administration had conflated the WMD and Al Qaeda threat as a single, undifferentiated threat, and that this policy "obscure[d] critical differences among rogue states, among terrorist organizations, and between rogue states and terrorist groups." Jeffrey Record, *Bounding the Global War on Terrorism* (Carlisle, PA: Strategic Studies Institute, 2003), v, 16.

24. The following paragraph is merely a summary of "The Ground War," to provide context for later analysis on its short-term effects in the initial counterinsurgency efforts. One of the best sources for further research into this conventional campaign is Gordon and Trainor's *Cobra II: The Inside Story of the Invasion and Occupation of Iraq*.

25. BC040, Battalion Commander, Interview by Benjamin Boardman and Richard Johnson, Fort Bragg, NC, 2 March 2011. This plan included an airhead and lodgment at Saddam International Airport on the outskirts of Baghdad, which was cancelled to enable the opening of a second front to the north by an airborne operation of the 173rd Airborne Brigade.

26. Chief among these measures was the prohibition from flying flags or displaying any other overt signs of foreign power within direct view of the Iraqi population.

27. For an excellent analysis of *jihad* within the Islamic law of warfare, see: Majid Khadduri, *War and Peace in the Law of Islam* (Baltimore, MD: Johns Hopkins Press, 1955), 51-136.

28. Lewis, *The Crisis of Islam*, xxiv-xxvi, xxxii, 63, 166.

29. Ahmed S. Hashim, "The Insurgency in Iraq," *Small Wars and Insurgencies* 14, no. 3 (August 2003): 3.

30. Carter Malkasian, "Counterinsurgency in Iraq: May 2003-January 2010," in *Counterinsurgency in Modern Warfare*, ed. Daniel Marston and Carter Malkasian (Oxford, UK: Osprey Publishing, 2010), 288.

31. Michael R. Gordon and Bernard Trainor, *Cobra II: The Inside Story of the Invasion and Occupation of Iraq* (New York: Pantheon Books, 2006), 580.

32. Thomas E. Ricks, *The Gamble* (New York: Penguin Press, 2009), 195.

33. Peter R. Mansoor, *Baghdad at Sunrise: A Brigade Commander's War in Iraq* (New Haven, CT: Yale University Press, 2008), 256. Due to their preponderance of forces in the counterinsurgency effort, all military organizations

and units in this chapter are US Army forces unless otherwise noted.

34. Hashim, "The Insurgency in Iraq," 5-6.

35. Carter Malkasian, "The Role of Perceptions and Political reform in Counterinsurgency: The Case of Western Iraq, 2004-2005," *Small Wars and Insurgencies* 17, no 3 (September 2006): 371.

36. Malkasian, "The Role of Perceptions and Political reform in Counterinsurgency," 380.

37. Gordon and Trainor, *Cobra II*, 358-359.

38. Gordon and Trainor, *Cobra II*, 434-435, 456.

39. Gordon and Trainor, *Cobra II*, 505.

40. Hashim, "The Insurgency in Iraq," 16.

41. Hashim, "The Insurgency in Iraq," 7-8.

42. Gordon and Trainor, *Cobra II*, 445.

43. Kilcullen, *The Accidental Guerrilla*, xviii-xix. This analysis is corroborated by the author's experience in the overwhelmingly Sunni region of Salah-ad-Din from 2006 to 2007, where most considered AQI to be a foreign force and irreverent of both tribal considerations and widely-held Islamic beliefs. The term *takfiri* was the common parlance to describe AQI as well as their affiliated groups.

44. Gordon and Trainor, *Cobra II*, 447.

45. Timothy S. McWilliams, *Al-Anbar Awakening: U.S. Marines and Counterinsurgency in Iraq, 2004-2009* (Quantico, VA: Marine Corps University Press, 2009), 86.

46. Hashim, "The Insurgency in Iraq," 9.

47. Matt Damon, *Rounders*, Directed by John Dahl, Santa Monica, CA: Miramax Films, 1998.

48. Gordon and Trainor, *Cobra II*, 466-467. This is evidenced in the inherently flawed effort for reorganizing Iraqi security forces, since it relied on a period of six months to analyze, weigh recommendations, and refine a separate plan. These six months simply did not exist. Gordon and Trainor attribute this analysis to Robert Perito, a civilian expert on peacekeeping operations at the United States Institute for Peace as he advised the Defense Policy Board.

49. Ricks, *Fiasco*, 59-60.

50. Ricks, *Fiasco*, 121.

51. Gordon and Trainor, *Cobra II*, 461. This quote was provided in the account of the Secretary of the Army Thomas White, in his description of Rumsfeld's efforts.

52. Tommy Franks, *American Soldier* (New York: Regan Books, 2004), 332-334.

53. Gordon and Trainor, *Cobra II*, 457, 461. Even though Arab countries withheld their contingents due to domestic political sensitivities, Rumsfeld

cancelled the planned deployments for two divisions of follow-on forces. Of these two divisions, the 1st Armored Division and the 1st Cavalry Division, only the 1st Armored Division would be turned back around to deploy in April 2003.

54. Ricks, *Fiasco*, 157.

55. Gordon and Trainor, *Cobra II*, 461, 502.

56. AA510, Former DivArty Commander, Interview by Travis Moliere and Jesse Stewart, Fort Leavenworth, KS, 4 November 2010. Emphasis added.

57. BF020 Civilian Adviser to MNF-I, Interview.

58. Daniel Marston, "Adaptation in the Field: The British Army's Difficult Campaign in Iraq," *Security Challenges* 6, No. 1 (Autumn 2010): 71. This notion was corroborated repeatedly in interviews, pointedly with a British General Officer during a previous interview. AA1012, British General Officer, Interview by Mike Dinesman, Ken Gleiman, Brian McCarthy, and Jesse Stewart, United Kingdom, 22 September 2010.

59. Author's experience in receiving training from British NCOs in Baghdad, 2003. The training was an excellent overview of lessons learned for patrolling techniques, but the scope was extremely limited.

60. Mansoor, *Baghdad at Sunrise*, 34, 357.

61. BD010, Field Grade Officer, Interview by Benjamin Boardman and Dustin Mitchell, Fort Knox, KY, 14 March 2011.

62. Gordon and Trainor, *Cobra II*, 586-590. The appendix contains the CPA order.

63. Gordon and Trainor, *Cobra II*, 564. This appendix contains the ORHA briefing slide concerning post-invasion considerations.

64. BF020, Civilian Adviser to MNF-I, Interview.

65. AA1013, British General Officer, Interview by Karsten Haake, Matt Marbella, Travis Molliere, and Carrie Przelski, United Kingdom, 23 September 2010.

66. BH030, Iraq Veterans Panel, Interview by Mark Battjes, Robert Green, Aaron Kaufman, and Dustin Mitchell, Washington, DC, 22 March 2011. A field grade officer on this panel was directly involved in the writing of that strategy.

67. BH030, Iraq Veterans Panel.

68. Gordon and Trainor, *Cobra II*, 463-464.

69. Gordon and Trainor, *Cobra II*, 574. This appendix is a copy of the original cable to London, titled "Iraq: What's Going Wrong."

70. BG040, Brigade Commander, Interview by Nathan Springer and Thomas Walton, Fort Stewart, GA, 15 March 2011.

71. Gordon and Trainor, *Cobra II*, 475. In an attempt to replace what the administration saw as ORHA's bias from having too many "arabists" on the staff, Bremer took leadership of the CPA despite never serving in a post in the Middle East before.

72. Ricks, *Fiasco*, 204-205; 209-210.

73. L. Paul Bremer, *My Year In Iraq* (New York: Simon and Schuster, 2006), 213.

74. BG040, Brigade Commander, Interview. The respondent was a battalion commander at the time with direct knowledge of this approach. Malkasian, "Counterinsurgency in Iraq: May 2003-January 2010," 289.

75. Gordon and Trainor, *Cobra II*, 452-453. The al-Jabouri tribe was one of the largest Sunni tribes in the Middle East, and had a reputation of criminal activity and nefarious dealings amongst other Sunni tribes. The *peshmerga* were the paramilitary force of the ethnic Kurdish community, widely regarded as the most capable (and the only controlled) indigenous fighting force in the aftermath of the ground war and De-Ba'athification.

76. Ricks, *Fiasco*, 229.

77. Ricks, *Fiasco*, 234.

78. BH020, Field Grade Officer, Interview by Mark Battjes, Ben Boardman, Robert Green, Richard Johnson, Aaron Kaufman, Dustin Mitchell, Nathan Springer, and Thomas Walton, Washington, DC, 21 March 2011.

79. BF020, Civilian Adviser to MNF-I, Interview.

80. BH020, Field Grade Officer, Interview.

81. Burton and Nagl, "Learning as We Go," 304, 306. Although an uncirculated campaign plan was in draft form, there was no plan to provide guidance and continuity in operations until Casey took command in 2004.

82. George Packer, "The Lesson of Tal Afar," *The New Yorker* 82, no. 8 (10 April, 2006), www.newyorker.com/archive/2006/04/10/ 060410fa_fact2 (accessed 13 May 2011). The quote, attributed to Colonel Thomas X. Hammes (who worked as a military officer in CPA from 2003-2004) is as follows: "Why is the 82nd hard-ass and the 101st so different?" Hammes asked. "Because Swannack sees it differently than Petraeus. But that's Sanchez's job. That's why you have a corps commander."

83. Ricks, *Fiasco*, 173.

84. Ricardo Sanchez, *Wiser in Battle: A Soldier's Story* (New York: Harpercollins, 2008), 444-446.

85. Burton and Nagl, "Learning as We Go," 304; Ricks, *Fiasco*, 173.

86. Ricks, *Fiasco*, 193.

87. Mansoor, *Baghdad at Sunrise*, 35, 228. Mansoor's BCT in East Baghdad operated with a lower force to population ratio (1:600) than the British in Northern Ireland (1:50) or even the NYPD (1:209). Additionally, he refers to HUMINT as "the coin of the realm in counterinsurgency" and cites it as his chief obstacle to success in 2003 and 2004.

88. Gordon and Trainor, *Cobra II*, 492.

89. Ricks, *Fiasco*, 213.

90. Ricks, *Fiasco*, 221.

91. Ricks, *Fiasco*, 261, 280.

92. Ricks, *Fiasco*, 326.

93. Bing West, *No True Glory: A Frontline Account of the Battle for Fallujah* (New York: Bantam Dell, 2005), 56; Ricks, *Fiasco*, 142.

94. Malkasian, "Counterinsurgency in Iraq: May 2003-January 2010," 289.

95. West, *No True Glory*, 49-62; Malkasian, "The Role of Perceptions and Political reform in Counterinsurgency: The Case of Western Iraq, 2004-2005," 373; Ricks, *Fiasco*, 332-333. Lieutenant General James T. Conway, commander of the I Marine Expeditionary Force, describes the decision in Ricks' account: "We felt like we had a method that we wanted to apply to Fallujah: that we ought to probably let the situation settle before we appeared to be attacking without revenge." The author's conversations with many small unit leaders in the 82nd Airborne Division in 2004 provide an alternate analysis; that the Marines had little tactical momentum at first since they failed to absorb the paratroopers' understanding of the complex local situation, exhibited relatively poor tactical discipline at first, and characterized the violence in Fallujah as insurgent activity rather than criminal activity with anecdotal foreign fighter presence. Additionally, many leaders at the company level and below left with the impression that the Marines had settled on a soft, indirect approach before assessing the situation in-country. In any case, this episode demonstrates the vulnerabilities and lack of continuity during repeated unit rotations in OIF.

96. Malkasian, "Counterinsurgency in Iraq: May 2003-January 2010," 291. While the Marines had planned on a partnered offensive with the 2nd Battalion of the New Iraqi Army, the unit deserted en masse before entering the battle. Only 70 Iraqis accompanied more than 2,000 Marines on the operation.

97. Anthony Cordesman, *Iraqi Security Forces: A Strategy For Success* (Westport, CT: Praeger Security International, 2006), 357.

98. Ricks, *Fiasco*, 343.

99. Ricks, *Fiasco*, 337, 358.

100. Mansoor, *Baghdad at Sunrise*, 353.

101. Malkasian, "Counterinsurgency in Iraq: May 2003-January 2010," 292.

102. Ricks, *Fiasco*, 362. Ricks' analysis cites the view of Major General Charles Swannack, who commanded the 82nd Airborne Division in 2003-2004 in OIF.

103. Cordesman, *Iraqi Security Forces*, 87-91.

104. B010, Field Grade Officer, Interview.

105. Malkasian, "Counterinsurgency in Iraq: May 2003-January 2010," 294; Ricks, *Fiasco*, 390.

106. Burton and Nagl, "Learning as We Go," 304-306.

107. Burton and Nagl, "Learning as We Go," 306. These best practices were eventually refined and published in *Military Review* in 2005. Refer to Kalev Sepp, "Best Practices in Counterinsurgency," *Military Review* (May-June 2005): 10.

108. Burton and Nagl, "Learning as We Go," 307; Ricks, *Fiasco*, 392.

109. Malkasian, "Counterinsurgency in Iraq: May 2003-January 2010," 294.

In oral histories, most OIF veterans will refer to MNSTC-I by the nickname "Min-Sticky" due to the unwieldy acronym's lack of vowels.

110. Ricks, *Fiasco*, 392. Ricks quotes Andrew Bacevich in this analysis of Sanchez's likely legacy as "the Westmoreland of Iraq."

111. Malkasian, "The Role of Perceptions and Political reform in Counterinsurgency: The Case of Western Iraq, 2004-2005," 375. The Sunni community largely boycotted the national elections to the point that only 2 percent to 5 percent of the population in Al Anbar Province voted. Since Kurd and Shia groups organized political structures in opposition to the Ba'ath Party (both in secret and in exile), they were able to mobilize support relatively quickly in contrast to the Sunni minority which never truly organized in a political sense. As a result, many Sunni areas considered themselves outside of the effective governance of the GoI, and tribal relationships maintained their influence.

112. Jay B. Baker, "Tal Afar 2005: Laying the Counterinsurgency Groundwork," *Army* (June 2009): 63-64. This account of the counterinsurgency effort in Tal Afar was written by the regimental surgeon, and is highly regarded by several other tactical leaders of that operation.

113. Malkasian, "Counterinsurgency in Iraq: May 2003-January 2010," 298.

114. Packer, "The Lesson of Tal Afar." Since much of the population had escaped and firepower was deliberately limited, collateral damage was nowhere near the scale seen in Fallujah in 2004. In addition to easing the transition with civic authorities, this enabled critical infrastructure projects to begin much sooner. One field grade officer with 3rd ACR commented that "[y]ou can come in, cordon off a city, and level it, à la Fallujah. Or you can come in, get to know the city, the culture, establish relationships with the people, and then you can go in and eliminate individuals instead of whole city blocks." Also, refer to Ricks, *Fiasco*, 422-423.

115. BH070, Iraqi Mayor, Interview by Mark Battjes and Robert Green, Washington, DC, 25 March 2011. This mayor had direct knowledge of the operations in Tal Afar.

116. Packer, "The Lesson of Tal Afar." This is reflected in a quote by one of the ACR's Squadron Commanders: "We tried to switch the argument from Sunni versus Shia, which was what the terrorists were trying to make the argument, to Iraqi versus *takfirin*."

117. BH070, Iraqi Mayor, Interview.

118. Baker, "Tal Afar 2005: Laying the Counterinsurgency Groundwork," 67. Unfortunately, when units from 1st Armored Division replaced 3rd ACR in 2006 they returned to a heavy reliance on mass detentions; attack levels rose once again. Burton and Nagl, "Learning as We Go," 309.

119. Malkasian, "Counterinsurgency in Iraq: May 2003-January 2010," 298.

120. BH040, Afghanistan Veterans Panel, Interview by Richard Johnson, Aaron Kaufman, Nathan Springer, and Thomas Walton, Washington, DC, 24

March 2011. A respondent on this panel had direct knowledge of operations in Al Qaim.

121. Malkasian, "Counterinsurgency in Iraq: May 2003-January 2010," 298-299.

122. James A. Russell, *Innovation, Transformation, and War: Counterinsurgency Operations in Anbar and Ninewa Provinces, Iraq, 2005-2007* (Stanford, CA: Stanford University Press, 2011), 74-76. Russell's conclusions in this case study were largely echoed during interview with BH040, Afghanistan Veterans Panel.

123. Malkasian, "Counterinsurgency in Iraq: May 2003-January 2010," 299. The term "MiTT" represents a Military Training Team, a blanket term to describe advisory teams with IA units at any echelon.

124. AA810, Battalion Commander, Interview by Ken Gleiman, Matt Marbella, Brian McCarthy, and Travis Molliere, Washington, DC, 13 September 2010.

125. Malkasian, "Counterinsurgency in Iraq: May 2003-January 2010," 300.

126. Ricks, *Fiasco*, 438.

127. Ricks, *The Gamble*, 14. Ricks cites President Bush's speech at Annapolis, MD.

128. BA010, Brigade Commander, Interview by Richard Johnson and Thomas Walton, Fort Leavenworth, KS, 22 February 2011.

129. BI020, Battle Group Commander, Interview by Aaron Kaufman and Thomas Walton, United Kingdom, 31 March 2011.

130. BI020, Battle Group Commander, Interview. One incongruence with this comparison is that SVN had a huge, popularly elected government and state-aligned local security forces after the Tet Offensive, and Iraq lacked both an effective central government and security forces. But in overall theater strategic terms, the comparison is apt.

131. BH030, Iraq Veterans Panel,, Interview.

132. Malkasian, "Counterinsurgency in Iraq: May 2003-January 2010," 301.

133. Ricks, *The Gamble*, 45.

134. Ricks, *The Gamble*, 173. Kilcullen, who would serve as a counterinsurgency adviser to Petraeus in Baghdad, relates that: "we did a counterintelligence assessment of an Iraqi army battalion in central Baghdad and found that every senior commander and staff were either JAM, doing criminal activity with JAM, or intimidated by JAM."

135. Kilcullen, *The Accidental Guerrilla*, 126.

136. Ricks, *The Gamble*, 45.

137. Kilcullen, *The Accidental Guerrilla*, 141-143. Kilcullen provides an in-depth analysis of this cycle of violence, and the efforts to interrupt it.

138. BH020, Field Grade Officer, Interview.

139. Malkasian, "Counterinsurgency in Iraq: May 2003-January 2010," 301.

140. Malkasian, "Counterinsurgency in Iraq: May 2003-January 2010," 302. Only 1,000 of the planned 4,000 Iraqi soldiers arrived in Baghdad.

141. Burton and Nagl, "Learning as We Go," 316.

142. Malkasian, "Counterinsurgency in Iraq: May 2003-January 2010," 302.

143. BC030, Battalion Commander; Interview by Benjamin Boardman and Richard Johnson, Fort Bragg, NC, 1 March 2011; Burton and Nagl, "Learning as We Go," 317.

144. Sean MacFarland and Neil Smith, "Anbar Awakens: The Tipping Point," *Military Review* (March-April 2008): 42; Malkasian, "Counterinsurgency in Iraq: May 2003-January 2010," 303. AQI viewed Ramadi as the future capital of its caliphate in Iraq and enjoyed relative freedom of movement in the area, making it almost exclusively denied terrain in the eyes of coalition forces.

145. McWilliams, *Al-Anbar Awakening*, 91.

146. BA020, Battalion Commander, Interview by Mark Battjes and Benjamin Boardman, Fort Leavenworth, KS, 23 February 2011. The respondent asserted that "there wasn't a strategy in Ramadi. What Colonel MacFarland really brought was a strategy. The Marines certainly didn't have a strategy. The Marines in my mind, if Ramadi was the center of the government and the division HQ moved from Ramadi to Fallujah, what message did this send?"

147. MacFarland and Smith, "Anbar Awakens," 43. One factor that enabled this was the leeway given to an Army unit serving under a Marine headquarters, corroborated in interview with BA010, Brigade Commander (the respondent has direct knowledge of these operations in Ramadi).

148. BA010, Brigade Commander, Interview. The unit had to refer to these as "substations" since a rigid plan from Baghdad had already templated locations for Iraqi Police stations in Ramadi without consulting the local leaders.

149. BA010, Brigade Commander, Interview; MacFarland and Smith, "Anbar Awakens: The Tipping Point," 43, 45. To strengthen the effect of integrating existing tribal forces into effective local security, additional militias and bodyguards for sheikhs were deputized as Provincial Auxiliary Iraqi Police. This had the added benefit of trickling money into the local economy through these fighters' wages

150. Panel discussion during US Army Command and General Staff College Art of War Scholars Seminar, Iraq Session, 3 February 2011, Fort Leavenworth, KS.

151. Burton and Nagl "Learning as We Go,", 313.

152. Marston, "Adaptation in the Field," 75.

153. AA810, Battalion Commander, Interview; Burton and Nagl, "Learning as We Go," 310.

154. Ricks, *The Gamble*, 24; Kilcullen, The Accidental Guerrilla, 119. In a reflection of the traditional Army expression that "doctrine is too important to be left to the doctrine writers," Petraeus assembled an international cast of academics

and practitioners to develop a presentation of ideas instead of a standardized, prescriptive field manual.

155. BA090, Brigade Commander, Interview by Mark Battjes and Benjamin Boardman, Fort Leavenworth, KS, 24 February 2011.

156. BA090, Brigade Commander, Interview.

157. BA010, Brigade Commander, Interview; BA090, Brigade Commander, Interview.

158. Kilcullen, *The Accidental Guerrilla*, 214.

159. BH030, Iraq Veterans Panel,, Interview.

160. Kilcullen, *The Accidental Guerrilla*, 126.

161. The Baker-Hamilton Commission, *Iraq Study Group Report: Gravel Edition* (Washington, DC: Filiquarian Publishing, 2006), 9, 55, 71, 72-76. To achieve this, the group's report advocated two simultaneous approaches to achieve a satisfactory strategic outcome and a complete withdrawal of the American military. The "external approach" consisted of using regional powers such as Syria and Iran in a diplomatic effort to encourage insurgents and disaffected groups to reconcile with the GoI. The "internal approach" made ISF assistance the primary mission of American forces, to facilitate a hasty withdrawal. Progress would enable a conditions-based withdrawal, as measured by milestones for the GoI in the areas of national reconciliation, security, and governance.

162 Frederick W. Kagan, *Choosing Victory: A Plan for Success in Iraq* (Washington, DC: American Enterprise Institute, 2006), 1. After vetting the concepts and operational feasibility of the plan with McMaster and some of his veterans of the Tal Afar campaign, it was refined by a council of colonels. AEI presented the concept to several congressmen, then President Bush reviewed that proposal.

163. Ricks, *The Gamble*, 95-97. Refer to Appendix B in *The Gamble* for the presentation to Odierno regarding Casey's operational approach to transition forces back from the cities into large FOBs, called the *Transition Bridging Strategy* dated December 2006.

164. David Ucko, *The New Counterinsurgency Era: Transforming the US Military for Modern Wars* (Washington DC: Georgetown University Press, 2009), 120-122. Since this redoubled effort and accelerated deployment of forces lacked domestic support, the US Congress tied many benchmarks to the authorization bill; the counterinsurgency effort would be tied directly to Prime Minister Nouri al-Maliki and the GoI's limited control.

165. BH030, Iraq Veterans Panel, Interview. By virtue of his position, the respondent had direct access to the decision process at MNF-I. Also, see Malkasian, "Counterinsurgency in Iraq: May 2003-January 2010," 304-305.

166. Ricks, The Gamble, 356. Refer to Appendix C in The Gamble for Odierno's inbrief to Petraeus dated 8 February 2007. The Baghdad belts were the urban centers, connecting river valleys, and infrastructure networks surrounding the metropolitan capital.

167. Ricks, *The Gamble*, 200-201. This quote is attributed to Captain Jay

Wink, an S-2 in southwest Baghdad at the time.

168. BF020, Civilian Adviser to MNF-I, Interview.

169. Kilcullen, *The Accidental Guerrilla*, 126-127.

170. Interview with BH020, Field Grade Officer, corroborated interview with BH030, Iraq Veterans Panel, in which a respondent added that the reintroduction of these 10,000 families became a focal point for higher headquarters in later stages of the campaign.

171. Kilcullen, *The Accidental Guerrilla*, 135. The number of JSSs that forces established is different in almost every account, since some units considered every patrol base a JSS, while others classified US-only bases or joint locations at the platoon level and below as Combat Outposts. For example, Malkasian reports in "Counterinsurgency in Iraq: May 2003-January 2010" that 50 JSSs were emplaced in Baghdad.

172. BF040, Battery Commander, Interview by Richard Johnson and Aaron Kaufman, Boston, MA, 14 March 2011. In accordance with the commander's guidance from both Odierno and Petraeus, units across Iraq began to emplace JSSs at key locations in their areas of operation. Some of these were abandoned within the last year by previous units as part of Casey's *Transition Bridging Strategy.*

173. Packer, "The Lesson of Tal Afar."

174. BA070, Battery Commander, Interview by Richard Johnson and Thomas Walton, Fort Leavenworth, KS, 24 February 2011. This commander operated in a JSS far afield of his parent headquarters, and stated that "in the counterinsurgency fight, you should be relying more and more on locals; it was all about CPT Asir's phone ringing."

175. Panel discussion during US Army Command and General Staff College Art of War Scholars Seminar, Iraq Session, 3 February 2011, Fort Leavenworth, KS.

176. BF030, Battery Commander, Interview by Richard Johnson and Aaron Kaufman, Boston, MA, 12 March 2011.

177. Panel discussion during US Army Command and General Staff College Art of War Scholars Seminar, Iraq Session, 3 February 2011, Fort Leavenworth, KS.

178. BF040, Battery Commander, Interview.

179. BC050, Battalion Commander, Interview by Benjamin Boardman and Richard Johnson, Fort Bragg, NC, 2 March 2011. The respondent was a battalion commander in Baghdad during this time.

180. Ricks, *The Gamble*, 174. Ricks quotes Lieutenant Colonel Dale Kuehl, a battalion commander in northwest Baghdad at the time.

181. BC050, Battalion Commander, Interview. Additionally, the respondent in interview with BA090 related a story of hiring local art students to paint the concrete walls, addressing a local grievance with local talent.

182. Panel discussion during US Army Command and General Staff College Art of War Scholars Seminar, Iraq Session, 3 February 2011, Fort Leavenworth, KS.

183. These local security forces were broadly referred to as "Concerned Local Citizens" forces in Baghdad and "Tribal Reconciliation Forces" in other areas; when higher commands and the GoI attempted to institutionalize the program it became known as the "Sons of Iraq."

184. BC030, Battalion Commander, Interview. The respondent's unit operated in the Baghdad belts during this "wave of moderation."

185. BA010, Brigade Commander, Interview.

186. BA070, Battery Commander, Battery Commander, Battery Commander, Interview.

187. BA010, Brigade Commander, Interview.

188. Ricks, *The Gamble*, 209. Ricks casts Gentile as one of these dissenters, quoting him as calling the effort "cash for cooperation."

189. Kilcullen, *The Accidental Guerrilla*, 165.

190. BC030, Battalion Commander, Interview.

191. BA070, Battery Commander, Interview.

192. BA090, Brigade Commander, Interview.

193. BF020, Civilian Adviser to MNF-I, Interview.

194. BH030, Iraq Veterans Panel,, Interview.

195. BF030, Battery Commander, Interview.

196. Ucko, *The New Counterinsurgency Era*, 124.

197. Ricks, *The Gamble*, 210

198. BA020, Battalion Commander, Interview.

199. Kilcullen, *The Accidental Guerrilla*, 173.

200. Ricks, *The Gamble*, 172. Ricks' account cites the experiences of C/1-26 IN in the Adamiyah neighborhood of Baghdad.

201. Kilcullen, *The Accidental Guerrilla*, 137.

202. Ucko, *The New Counterinsurgency Era*, 123.

203. Ricks, *The Gamble*, 190. Ricks quotes Major General Joseph Fil, the commander of MND-B during that time.

204. Ricks, *The Gamble*, 177; Ucko, *The New Counterinsurgency Era*, 129. In *The Gamble,* Ricks presents this as an unconfirmed rumor that Petraeus "flatly denied." Ucko's two cited sources are *Newsweek* articles which rely on interviews with Petraeus and Sadr officials in Baghdad: Babak Dehghanpisheh, "The Great Moqtada Makeover," *Newsweek* (19 January 2008), www.newsweek.com/2008/01/19/the-great-moqtada-makeover.html (accessed 8 June 2011); Rod Norland, "A Radical Cleric Gets Religion," *Newsweek* (10 November 2007), www.newsweek.com/2007/11/10/a-radical-cleric-gets-religion.html (accessed 8 June 2011). Norland's articles states that "[t]he general's spokesman, Col. Steven Boylan, qualified that assertion, explaining that while Petraeus has not met with Sadr, 'the command has indeed had direct engagements with some of his people

within the [Sadr] organization ... to assist with reconciliation efforts.' Boylan also says the military 'applauded' Sadr's ceasefire."

205. Ricks, *The Gamble*, 177. Kilcullen's use of the term "we" seems to give further weight to Ucko's assertion of Petraeus' discussion with the Sadrists, since Kilcullen was his counterinsurgency adviser.

206. Ucko, *The New Counterinsurgency Era*, 130-131.

207. Marston, "Adaptation in the Field," 76. The fact that the British did not participate at the Taji COIN Center until 2008 is ironic because its structure was based on the British Army's Far East Land Forces Training Centre described in Chapter Three.

208. BA010, Brigade Commander, Interview.

209. BC050, Battalion Commander, Interview.

210. BH020, Field Grade Officer, Interview.

211. BA010, Brigade Commander, Interview.

212. BA010, Brigade Commander. Interview.

213. BC030, Battalion Commander, Interview.

214. AA509, Division Commander, Interview by Ken Gleiman and Karsten Haake, Fort Leavenworth, KS, 25 August 2010.

215. BI020, Battle Group Commander, Interview. The respondent estimated that a third could be released immediately in 2007, a third could use short-term rehabilitation, and the remaining third would have to be held for further determination. In all, he felt that as little as 2,000 detainees could never be reconciled to the GoI, a seemingly manageable number.

216. BF040, Battery Commander, Interview.

217. BI290, Battery Commander, Interview by Richard Johnson, United Kingdom, 1 April 2011.

218. J. K. Wither, "Basra's Not Belfast," *Small Wars and Insurgencies* 20, no. 3 and 4 (September-December 2009): 622.

219. BI020, Battle Group Commander, Interview. BI110, Battalion Commander, Interview, indicated that this was due to the character of the region; southeastern Iraq was always a different environment than Baghdad and other cities of mixed populations due to the completely Shia Arab demographics of the area. Without concerns of communal violence, it was somewhat of an economy of effort when viewed against the entire scope of OIF. BI100, Senior Army Officer, Interview by Mark Battjes, Richard Johnson, Aaron Kaufman, and Dustin Mitchell, United Kingdom, 4 April 2011.

220. BI100, Senior Army Officer,, Interview.

221. Malkasian, "Counterinsurgency in Iraq: May 2003-January 2010," 306.

222. Wither, "Basra's Not Belfast," 615. During interview with BI010, the respondent (a senior British Army officer) summarized this in the terms of OIF within the context of international considerations: "The difficulty is that British politicians simply didn't want to be in Basra. It was a bad place, a bad war, and

we had Afghanistan to focus on. That was the good war. How things changed." BI010, Senior British Officer, Interview by Mark Battjes, Benjamin Boardman, Robert Green, Richard Johnson, Aaron Kaufman, Dustin Mitchell, and Nathan Springer, United Kingdom, 29 March 2011.

223. Malkasian, "Counterinsurgency in Iraq: May 2003-January 2010," 306.

224. Wither, "Basra's Not Belfast," 622.

225. Malkasian, "Counterinsurgency in Iraq: May 2003-January 2010," 306-307. BI010, Senior British Officer, provided a very similar analysis. BH030, Iraq Veterans Panel, related that the local IA commander, General Mohan, brokered the deal which was supposed to end with the British withdrawing from the city, no indirect fire attacks on the British FOB as they overwatched events, and a pacified Basra with effective local governance. When British forces left the city in September, they lost more than 100 HUMINT sources to assassination or intimidation

226. BI100, Senior Army Officer,, Interview. The total lack of British ISTAR assets after being re-allocated to Afghanistan was also cited as a critical shortcoming during BH030, Iraq Veterans Panel.

227. Malkasian, "Counterinsurgency in Iraq: May 2003-January 2010," 307. JAM's resumption of rocket attacks were also mentioned in BI010, Senior British Officer, Interview; and BH030, Iraq Veterans Panel,, Interview.

228. BI280, Commander's Panel, Interview by Richard Johnson, United Kingdom, 1 April 2011.

229. BI010, Senior British Officer, Interview.

230. BI100, Senior Army Officer,, Interview. This senior British Army officer attributed the successful operation to the MNC-I tactical forward headquarters' ability to coordinate the requisite ISF, bandwidth, intelligence, and fire support.

231. BC010, Field Grade Officer, , Interview.

232. BI100, Senior Army Officer,, Interview.

233. Malkasian, "Counterinsurgency in Iraq: May 2003-January 2010," 308.

234. Marston, "Adaptation in the Field," 75; Russell, 203.

235. BI110, Battalion Commander, Interview by Mark Battjes, Richard Johnson, and Dustin Mitchell, United Kingdom, 8 April 2011.

236. BH030, Iraq Veterans Panel,, Interview.

237. BI110, Interview. The respondent referred to the Permanent Joint Headquarters (PJHQ) in Northwood, UK.

238. Marston, "Adaptation in the Field," 81. Most Americans with direct experience in Basra felt that British soldiers and tactical leaders did an outstanding job, but were hamstrung by these artificial strategic and political considerations; Interview BH030 mentioned this directly.

239. BH030, Iraq Veterans Panel,,Interview. The respondent related this story and attributed the successful negotiations to Mansoor's early involvement in the process while he served as Petraeus' executive officer. Charles Levinson, "An Architect of U.S. Strategy Waits to Pop Cork," *Wall Street Journal* (27 August

10), online.wsj.com/article/ SB1000142405274870412560457544917125367744 4.html (accessed 20 May 2011).

240. BI290, Battery Commander, Interview.

241. Kilcullen, *The Accidental Guerrilla*, 161.

242. BF020, Civilian Adviser to MNF-I, Interview.

243. Ricks, *The Gamble*, 295.

244. BE080, Battalion Commander, Interview by Robert Green and Aaron Kaufman, Fort Irwin, CA, 7 March 2011.

245. AA810, Battalion Commander, Interview.

246. BH030, Iraq Veterans Panel, Interview. At the time, the respondent served as a battalion commander.

247. BI290, Battery Commander, Interview.

248. Malkasian, "Counterinsurgency in Iraq: May 2003-January 2010," 309. Efforts to turn over locations in Baghdad were hampered by political deals, competition, and corruption within the ISF, leading many locations to remain under American control well after forces were supposedly out of the cities. The largest JSS in Eastern Baghdad at the time of the security transition did not transfer authority to the 1st Federal Police Division until November.

249. BI290, Battery Commander, Interview.

250. BF020, Civilian Adviser to MNF-I, Interview.

251. Michael Christie, "Iraqi Civilian Deaths Drop to Lowest Level of War," *Reuters* (30 November, 2009), www.reuters.com/article/2009/11/30/us-iraq-toll-idUSTRESAT3ZE20091130 (accessed 20 May 2011).

252. BG020, Brigade Commander, Interview by Mark Battjes and Thomas Walton, Fort Stewart, GA, 14 March 2011. The AAB was a BCT that was supplemented with 48 field grade officers to give the unit an ability to resource 24 Transition Teams.

253. AA1009, Retired General Officer, Interview by Ken Gleiman, Brian McCarthy, Travis Molliere, and Carrie Przelski, United Kingdom, 4 October 2010. The respondent cites Aden as an example when he states that "[w]e really made a mistake in Aden by telling people when we were going to leave. It became clear that we were going to lose because we told everyone."

254. Emma Sky, "Iraq, From Surge to Sovereignty," *Foreign Affairs* (March-April 2011), www.foreignaffairs.com/articles/67481/emma-sky/iraq-from-surge-to-sovereignty (accessed 20 May 2011). Sky served as a political advisor to Odierno in Baghdad.

255. Robert M. Gates, *Request to Change the Name of Operation IRAQI FREEDOM to Operation NEW DAWN* (2010).

256. United States Forces Iraq, "Operation New Dawn/United States Forces Iraq," www.usf-iraq.com/new-face-of-iraq/operation-new-dawn (accessed 19 May 2011).

257. BA090, Brigade Commander, Interview.

258. Ricks, *The Gamble*, 129.

259. BF020, Civilian Adviser to MNF-I, Interview. *Team America: World Police* was a satirical movie released in 2004 which lampooned an elite team of marionette operatives that travelled the world creating more collateral damage than success. Upon release it immediately caught on with many soldiers in Iraq (partially due to the directors' previous work with the *South Park* television series), and units with a reputation for ridiculous collateral damage or complete ignorance to local customs were often times derided with the nickname "Team America."

260. Robert Scales, "Artillery's Failings in the Iraq War," *Armed Forces Journal* (November 2003): 44. Scales' article may be misleading within the context of this chapter; his analysis that the artillery branch must improve range and accuracy was written in the immediate reflections of the ground war, and only addresses the use of artillery on a conventional, albeit nonlinear, battlefield. For primary accounts illustrating the use of artillery during the ground war, see Thomas G. Torrance and Noel T. Nicolle, "Observations From Iraq: The 3rd Div Arty in OIF," *Field Artillery* (July-August 2003): 30-35; Patrick J. Sweeney and Jason C. Montgomery, "Iraq: 101st Division LNO in the V Corp FECC," *Field Artillery* (July-August 2003): 40-44.

261. Patricia Slayden Hollis, "Division Operations Across the Spectrum – Combat to SOSO in Iraq: Interview with Major General Raymond T. Odierno, CG of 4th ID in OIF," *Field Artillery* (March-June 2004): 10. Odierno described proactive counterfire as follows: "[t]he TUAV saw the mortarmen, and before the enemy could shoot the mortars, 3-16 FA destroyed them. Now that's proactive."

262. Ricks, *Fiasco*, 234. Ricks cites the *Field Artillery* interview noted above with Odierno's idea of proactive counterfire, but erroneously conflates that with his description of H&I fires and leaves out Odierno's actual description. Ricks continues with the unreferenced claim that Odierno's "assertion is at odds with the great body of successful counterinsurgency practice, which holds that firepower should be as restrained as possible, which is difficult to do with the long-range, indirect fire of artillery." It is important to note that Odierno was a career field artilleryman before becoming a General Officer, and intimately understood the tactical and technical considerations for employing artillery.

263. A reflection of this is the case of 2-3 FA from the 1st Armored Division. Deployed from 2003 to 2004, the battalion's Survey Platoon was featured on the cover of *Time* magazine to represent the American soldier as the 2003 Person of the Year. It was a fitting selection, since their adaptive methods of patrol operations in Adamiyah represented the effort of the artillery branch (and of the army as a whole) at the time to find its role in the counterinsurgency effort. Romesh Ratnesar and Michael Weisskopf, "Person of The Year 2003: Portrait Of A Platoon," *Time* (29 December 2003); Mansoor, 216-218.

264. BA020, Battalion Commander, Interview described the advantages of GMLRS within an urban environment; BA010, Brigade Commander, Interview also described their use in an urban environment. In the author's separate conversation with a former MLRS Battery Commander during OIF, he confirmed that all rockets fired were GMLRS, beginning in late 2005.

265. Ricks, *Fiasco*, 314.

266. Ricks, *Fiasco*, 319.

267. Ricks, *Fiasco*, 402. Accounts from the battle indicate that they could have used even more artillery. Since most fighting was generally at a range of 200 meters, other sources of HE such as man-portable rockets, grenades and light mortars were more responsive in the urban terrain and still sufficient.

268. BH070, Iraqi Mayor, Interview. Significantly, artillery fires were eventually used in this operation, but on a very limited basis.

269. BA010, Brigade Commander, Interview. Ramadi was formerly a large artillery training center for the old Iraqi Army, so there was a large number of well-trained mortar and rocket teams in the area. To combat this, counterfire drills were refined to return fires within 60 seconds. For additional background information on the employment of integrated counterfire operations, see J. O. Evans, *Counterfire Requirements in an Insurgency* (Quantico,VA: United States Marine Corps Command and Staff College, 2006).

270. MacFarland and Smith, "Anbar Awakens: The Tipping Point," 50.

271. BA040, Brigade Commander, Interview describes using artillery fires to continually suppress an area, and integrating direct fires from tanks to complement terrain denial in some areas. These operations were integrated with his own ability to engage local leaders regarding that site, and by integrating combat operations and interpersonal relationships he was able to affect change. Refer to BA040, Brigade Commander, Interview by Aaron Kaufman and Dustin Mitchell, 23 February, 2011, Fort Leavenworth, KS.

272. BE060, Brigade Commander. Interview by Mark Battjes and Thomas Walton, Fort Irwin, CA, 9 March 2011.

273. BA020, Battalion Commander, Interview.

274. BH020, Field Grade Officer, Interview. This BCT was eventually able to stand up a firing element of 155mm M109A6s to provide fires for their area of operations, which eventually expanded to cover most of the Westside of Baghdad.

275. BA090, Brigade Commander. Interview by Mark Battjes and Benjamin Boardman, Fort Leavenworth, KS, 24 February 2011.

276. BB010, Battalion Commander. Interview by Mark Battjes and Nathan Springer, Fort Bliss, TX, 2 March 2011.

277. What follows is a description of several anecdotal accounts from five commanders operating in that area. What is striking is the similarity that developed in relative isolation from one another, in areas of common threat, terrain, and urban-rural demographics. Due to the limits of research, these instances are evidence of a trend in the evolution of artillery units' use in OIF, not a revolution.

278. Hollis, "Division Operations Across the Spectrum – Combat to SOSO in Iraq," 11. Odierno described his division's operations as such: "Every one of my artillery battalions owned its own battlespace. My FA battalions were just like my maneuver battalions. And every one had Bradleys and tanks working for them. That's the kind of flexibility we need as we look to the future." For an in-depth example of one battalion's split operations early in OIF, see Richard M. Cabrey

and Douglas M. Thomas, "1-5 FA in OIF II: Maintaining FA Competencies While Deployed," *Field Artillery* (January-February 2007): 15-19.

279. BA040, Brigade Commander, Interview.

280. BC030, Battalion Commander, Interview.

281. BG100, Brigade Commander, Interview by Mark Battjes and Nathan Springer, Fort Stewart, GA, 16 March 2011.

282. BA070, Battery Commander, Interview.

283. BI290, Battery Commander, Interview.

284. BI280, Commander's Panel, Interview.

285. BI280, Commander's Panel, Interview.

286. BI280, Commander's Panel, Interview.

287. BI280, Commander's Panel, Interview.

288. BE040, Transition Team Leader, Interview by Mark Battjes and Thomas Walton, Fort Irwin, CA, 9 March 2011.

289. BG020, Brigade Commander, Interview.

290. BI290, Battery Commander, Interview. The respondent related an anecdote of one unit's use of an Excalibur mission to destroy a suspected IED trigger location, with local leaders gathered to demonstrate that the ISF still had American artillery backing.

291. AA610, Platoon Leader and FDO, Interview by Travis Molliere and Carrie Przelski, Fort Bragg, NC, 18 August 2010.

292. BC030, Battalion Commander, Interview.

293. BA060, Battalion Commander, Interview by Robert Green and Nathan Springer, Fort Leavenworth, KS, 23 February, 2011. The respondent served as a battalion commander in OIF.

294. For instance, both examples of adroit "company-level commanders" co-opting local insurgent actors in the preceding section were field artillery officers commanding converted maneuver force batteries.

295. BG100, Brigade Commander, Interview.

296. BC050, Battalion Commander, Interview.

297. BH030, Iraq Veterans Panel, Interview. The respondent described this initiative due to his staff duties at the time, and judged that "We really didn't do [the Iraqis] any favors" in delaying their artillery training.

298. MacFarland and Smith, "Anbar Awakens: The Tipping Point," 45.

299. BA010, Brigade Commander, Interview. While not the commander of the artillery unit that ran this training center, the respondent had direct knowledge of it by virtue of his position at the time.

300. BA060, Battalion Commander, Interview.

301. The following section does not intend to explore every instance of artillery units serving in a nonstandard role during OIF; that would require an

entire volume of work. What follows is a brief discussion on some of the most common uses of artillery units during the counterinsurgency effort.

302. Lynne Garcia, *Interview with Colonel Frank Hull* (Fort Leavenworth, KS: Combat Studies Institute, 2005), 5. Hull was the DivArty Commander for 1st Armored Division during their deployment to OIF in May 2003.

303. BG040, Brigade Commander, Interview. The respondent provides an example of the "Team Village" concept.

304. AA510, Fomer DivArty Commander, Interview. The respondent describes using an attached artillery unit in this manner.

305. Garcia, *Interview with Colonel Frank Hull*, 4, 8.

306. BD020, Commander, Interview by Benjamin Boardman and Dustin Mitchell, Fort Knox, KY, 14 March 2011. Of note, this requirement was so large that the artillery battalion required additional augmentation from infantrymen and cavalry troopers.

307. BB030, Brigade Commander, Interview by Mark Battjes and Nathan Springer, Fort Bliss, TX, 3 March 2011. The strain of this requirement was also discussed in interview BC030, Battalion Commander, as well as the evident relief as the stateside combat training centers de-emphasized this mission in 2011 when their focus returned to full spectrum operations.

308. Peter Chiarelli and Patrick Michaelis, "The Requirements for Full-Spectrum Operations," *Military Review* 85, no. 4 (July-August 2005): 16.

309. Malkasian, "The Role of Perceptions and Political reform in Counterinsurgency: The Case of Western Iraq, 2004-2005," 387.

310. AA509, Division Commander, Interview.

311. BF020, Civilian Adviser to MNF-I, Interview.

312. BC010, Field Grade Officer, Interview by Robert Green and Aaron Kaufman, Fort Bragg, NC, 1 March 2011.

313. Ucko, *The New Counterinsurgency Era*, 125; Malkasian, "The Role of Perceptions and Political reform in Counterinsurgency: The Case of Western Iraq, 2004-2005," 381.

314. Gordon and Trainor, *Cobra II*, 469.

315. Bremer, *My Year In Iraq*, 220.

316. Ricks, *Fiasco*, 371. Ricks quotes Hammes as follows: "[t]he contractor was hired to protect the principal. He had no stake in pacifying the country. Therefore, they often ran Iraqis off the roads, reconned by fire, and generally treated locals as expendable." For a discussion on the overall usage of armed contractors in OIF, refer to Sarah K. Cotton et al., *Hired Guns: Views About Armed Contractors in Operation Iraqi Freedom* (Santa Monica, CA: RAND, 2010), 63-66.

317. BC030, Battalion Commander, Interview corroborated by the author's experience securing PRT ground movements during OIF 06-08 and serving on a BCT staff with an embedded PRT during OIF 08-09.

318. BF020, Civilian Adviser to MNF-I, Interview.

319. This notion was expressed in multiple interviews, with a very nuanced and mature understanding in AA810, Battalion Commander, Interview; AA509, Division Commander, Interview; and BF020, Civilian Adviser to MNF-I, Interview. The most commonly cited issues were the special operations forces' lack of concern for both collateral damage and damage to the existing relationships in the area of operations.

320. AA402, Special Forces Group Commander, Interview by Brian McCarthy and Jesse Stewart, Fort Lewis, WA, 31 August 2010. This commander's irritation of hearing these repeated accusations was evident: "It chaps my ass to no f---ing end when we go downrange in Iraq or Afghanistan and some conventional guy says 'all SF guys want to do is kick down doors.'"

321. AA402, Special Forces Group Commander, Interview.

322. BI020, Battle Group Commander, Interview.

323. Panel discussion during US Army Command and General Staff College Art of War Scholars Seminar, Iraq Session, 3 February 2011, Fort Leavenworth, KS.

324. BA040, Brigade Commander, Interview.

325. BC030, Battalion Commander, Interview.

326. BA040, Brigade Commander, Interview.

327. Chiarelli and Michaelis, "The Requirements for Full-Spectrum Operations," 16.

328. BA070, Battery Commander, Interview.

329. This is especially true for JAM and the Iraqi Police, since there was a large overlap of membership in those two groups.

330. BF040, Battery Commander, Interview.

331. Kilcullen, *The Accidental Guerrilla*, 145.

332. Craig Collier, "Now That We're Leaving Iraq, What Did We Learn?" *Military Review* (September-October 2010): 88-89.

333. Collier, "Now That We're Leaving Iraq, What Did We Learn?", 89.

334. Malkasian,"The Role of Perceptions and Political reform in Counterinsurgency: The Case of Western Iraq, 2004-2005," 381-382.

335. The military's need for organic fires in all forms of warfare is best captured in Robert F. Barry, "Why Organic Fires?" *Field Artillery* (March-June 2004): 13-18.

336. John Brock, "We Cannot Take Your Call for Fire Right Now: Does the Global War on Terror Signal the Demise of the Field Artillery?" (Monograph, School of Advanced Military Studies, Fort Leavenworth, KS, 2006), 60-61. This SAMS monograph concludes that the artillery was neither a "dying branch" nor in danger of losing relevance due to issues such as army transformation, the decreased emphasis on high-intensity conventional capabilities and the cancellation of the Crusader howitzer program.

337. Kenneth Cosgriff and Richard Johnson, "Organizational Resistance to

Change in the Airborne Field Artillery" (Master's thesis, Webster University, 2006), 45-48. The primary sources of this resistance to change were the uncertainty of the future role and command relationships for fire supporters, and the absence of a definitive and enduring set of artillery senior leadership to act as the "change agent."

338. Concurrently, most Field Artillery Brigades began the conversion to Fires Brigades, which added several enablers and support units to the existing brigade-level structure for general support fires.

339. BD070, Field Grade Officer, Interview by Benjamin Boardman and Dustin Mitchell, 16 March 2011, Fort Knox, KY.

340. Ricks, *Fiasco*, 316. Ricks quotes a battalion commander in his assessment of operations in Fallujah in 2003 regarding collateral damage.

341. Ricks, *The Gamble*, 168.

342. BC030, Battalion Commander, Interview.

343. BI280, Commander's Panel, Interview.

344. BA010, Brigade Commander, Interview. The respondent stated that it was actually easier for him to authorize dropping a 500-lb JDAM than to drop a leaflet.

345. Joshua D. Mitchell, "Afghanistan: Firing Artillery Accurately with Air Force Met Support," *Field Artillery* (January-February 2003): 38.

346. BF040, Battery Commander, Interview.

347. BH030, Iraq Veterans Panel, Interview.

348. BE060, Brigade Commander, Interview.

349. BB010, Battalion Commander, Interview.

350. BA060, Battalion Commander, Interview.

351. BA040, Brigade Commander, Interview.

352. Sean MacFarland, Michael Shields, and Jeffrey Snow, *The King and I: The Impending Crisis in the Field Artillery's Ability to Provide Fire Support to Maneuver Commanders*, 1. This unpublished white paper immediately made the rounds of most artillery units. Tellingly, few intellectually curious artillery officers disagreed with the actual issues or recommendations as they were presented in the paper.

353. MacFarland, Shields, and Snow, *The King and I*, 2.

354. Panel discussion during US Army Command and General Staff College Art of War Scholars Seminar, Iraq Session, 3 February 2011, Fort Leavenworth, KS. The importance of freeing artillery leaders from this mindset was also stressed by a battery commander in BI290, Battery Commander, , Interview: "It can be pared down even further to 'what is the role of leadership?' No matter the task or mission at hand, you need good leaders. As far as COIN goes, the importance of leadership is to understand the environment you're working in."

355. In BF040, Battery Commander, Interview, the respondent cited his previous experience as a Battalion Fire Support Officer as his key to success in OIF.

356. BC050, Battalion Commander, Interview.

357. BH020, Field Grade Officer, Interview.

358. Of the 39 oral history interviews with commanders in which the role of artillery units in OIF was discussed at length, not a single one indicated that artillery leaders were found lacking in this demanding environment. Although this is not a definitive measure, it is indicative of a trend.

359. BD070, Field Grade Officer, Interview.

360. BB030, Brigade Commander, Interview.

361. AA610, Platoon Leader and FDO, Interview.

362. BC030, Battalion Commander, Interview.

363. BI290, Battery Commander, Interview. An EXEVAL is an External Evaluation, in this case a 48-72 hour test of proficiency on leader certifications and unit-level collective tasks.

364. Panel discussion during US Army Command and General Staff College Art of War Scholars Seminar, Iraq Session, 3 February 2011, Fort Leavenworth, KS.

365. AA610, Platoon Leader and FDO, Interview.

366. BC060, Battalion Commander, Interview by Ben Boardman and Richard Johnson, Fort Bragg, NC, 3 March 2011.

367. BF040, Battery Commander, Interview; Cabrey and Thomas, 17.

368. BI290, Battery Commander, Interview. This battery commander described the effort to regain firing capabilities in OIF through a short train-up between combat patrols, in only four weeks: "Overall, it was a success. Did we train to an ARTEP standard or train to the level where they could certify? Absolutely not. But there wasn't enough time . . . that was the first time the unit fired in over two years."

369. BC050, Battalion Commander, Interview.

370. BC040, Battalion Commander, Interview.

371. BH020, Field Grade Officer, Interview. This commander referred to the practice of augmenting artillery and armor units as "more alchemy than math."

372. BC030, Battalion Commander, Interview.

373. BC050, Battalion Commander, Interview.

374. BC030, Battalion Commander, Interview.

375. BF030, Battery Commander, Interview.

376. BA060, Battalion Commander, Interview. The respondent attributed this to a conversation he had with Petraeus prompted by *The King and I* white paper, in which he asserted that the artillery's newfound competencies in areas such as decentralized operations, leader training, and the ability to coordinate and synchronize more assets than ever before will enable the branch to regain its core competency of delivering indirect fires to pre-OIF standards.

377. This is based on the author's attempts to review lessons learned from

two Army repositories. The FA Lessons Learned site (www.us.army.mil/suite/kc/5480861) lacked any topical or unit after action reviews prior to 2009, although "FA Nonstandard missions" is the fourth priority for collection. The site is a good resource for MOS-specific tasks which will enable artillery retraining in the future, but there is little to no information regarding artillery units in counterinsurgency. The Center for Army Lessons Learned site (call2.army.mil/default.aspx) was a similar disappointment, with no usable content in the top 50 search results of "artillery + counterinsurgency," other than five articles from FA *Journal* and *Military Review*, plus one account from a chaplain's assistant in an artillery battalion in OIF. Counterintuitively, the author later found links to a limited number of CALL products within the "Pubs" section on the Fires Knowledge Network. Both sites were accessed on 13 May 2011. Lessons learned that are available on classified networks are beyond the purview of this study, but are only readily accessible to those leaders already in theater.

378. Kilcullen, *The Accidental Guerrilla*, 152.

Chapter 6
Conclusions

> The artillery community is on the horns of a dilemma: where does it fit, where does it make its biggest contribution? We have to be careful. In Afghanistan in 2011, we are not worried about the Taliban massing a battalion-six on us with some type of weapon system; our next adversary may. We can't leave the artillery at home. It is part of the combined arms fight.
>
> — Former Battalion Commander in OIF
> Interviewed 23 February 2011

This study began by presenting artillery as the counterinsurgent's biggest stick within the metaphor of a counterinsurgency effort that leverages both attractive "carrots" and coercive "sticks." This metaphor does not judge the inherent effectiveness of artillery's use, or imply that there are universal principles which govern its successful employment. It simply highlights the fact that successful counterinsurgencies are historically a pragmatic mix of these two approaches within a politically distributive outcome.

It is possible to isolate specific qualities in a given counterinsurgency effort, but only with a nuanced understanding of the campaign and its environment. Specific qualities for this study are comparable after the differences in the campaigns and environments of the Malayan Emergency, the American experience in Vietnam, and OIF are considered. Each counterinsurgency has its own unique characteristics, as does every artillery unit engaged in the effort. As such, overly prescriptive practices that are distilled from selected historical instances or hypothetical models hold a limited value. They are only needed in the absence of leaders who can understand their environment, identify a feasible solution, and adapt to its demands.

Accordingly, this study used historical case studies with varying elements in an attempt to isolate the broadly descriptive fundamentals which foster a sound employment of artillery units. But not even the deepest understanding of these fundamentals will ensure success in counterinsurgent warfare; even a merely adequate understanding of the population and their grievances will go much further to that end. The following factors and fundamentals should provide tactical leaders with a deeper understanding and a starting point in the employment of artillery units during a counterinsurgency, but not a solution. The absence of a definitive solution should not imply that there is no way to improve and prepare an army for future wars among the population; there is. This improvement and preparation lies within the leadership of both the unified counterinsurgency effort and the tactical leaders who will command soldiers on the ground.

The Factors of Employing Artillery Units in Counterinsurgency

Each of the four factors influencing the employment of artillery units may be relatively strong or weak in different counterinsurgency environments. In some cases, one factor may not exhibit a discernible influence due to the comparatively strong weight of the other three factors. Counterinsurgencies in Malaya, Vietnam, and Iraq adequately illustrated the factors of employing them: the requirement for indirect fires, constraints and limitations on indirect fires, the counterinsurgency effort's organization, and the conversion cost in adapting to nonstandard roles.

Requirements for Indirect Fires

The counterinsurgency effort's requirements for indirect fires is the primary factor which affects the employment of artillery units, but it is only one factor.[1] The OIF case study shows this through the use of split units during operations in the Baghdad belts area. American BCTs required indirect fires, but only to the extent that a single firing element had to be established in many cases. This gave those BCTs an additional headquarters and the combat power they needed to secure the population. The Malayan Emergency also exhibits this effect. Since RA batteries could effectively disperse to support sequential operations, only a limited number of them had to deploy. The Vietnam case study illustrates the strongest effect of this factor to the point that it precluded the use of artillery units in any other capacity. Coverage and responsiveness were definitely great capabilities to mitigate risk in the face of an enemy that could periodically mass forces. But MACV carried these qualities to an irrational end, seeking to cover every important area in SVN with responsive fires. With more artillery battalions than infantry battalions in a time of political limits on additional troop levels, the counterinsurgency effort suffered due to an inadequate force composition which reflected the ill-framed approach.[2]

Constraints and Limitations on Indirect Fires

The constraints and limitations on indirect fires within the counterinsurgency effort is the second factor to affect the employment of artillery units.[3] Different environments, demographics, and arrays of forces lead to different limitations, but constraints are generally comparable. Among these constraints, the issue of collateral damage in a counterinsurgency takes on a great institutional importance since it is specifically constrained to avoid tactically deleterious second- and third-order effects. However, physical limitations often exert the greatest force on a counterinsurgency effort, since they cannot always be overcome due to the immutable nature of some limits. Malaya was one such instance of this phenomenon. The physical limitations on observation and effective lethal fires in some of the thickest jungle in the world was simply too much to overcome with a technological or organizational approach. Since the effects of single-round fire missions for H&I and flushing fires had almost the same effect as a battery of massed fires, it allowed artillery units to disperse in extremely small detachments. Although Vietnam had a similar effect due to the limits of terrain, the main expression of this factor is seen in the emerging constraints of collateral damage concerns. In Vietnam, this was due to the extremely high volume of fires combined with the use of unobserved fires. In OIF, collateral damage was a similarly strong factor, but it was due to insurgents' presence in heavily developed or urbanized areas rather than a comparatively intense rate of indirect fire.

Counterinsurgent Force Organization

The organization of the counterinsurgency effort's military force also has a strong effect on the employment of artillery units. This factor manifests itself in the "pull" into nonstandard roles when there are insufficient maneuver forces to secure a population or area, and a "push" into those roles if other sources of indirect fires are effective and sufficient.[4] In OIF and the Malayan Emergency, this factor was considerably strong. In OIF, the "pull" for direct support artillery units to engage in security and other nonstandard mission profiles was sufficient to employ them in this manner more than any other counterinsurgency in modern history.[5] The "push" factor was just strong enough to enable this. Conversely, the same strong forces were at hand in Malaya which prompted the 26th RA to conduct nonstandard security missions before The Emergency was even declared, then this intense demand decreased due to the requisite amount of infantry units arriving in Malaya and improved local security forces. Only then did RA field batteries return to providing indirect fires, since there was virtually no "push" force from air assets.[6] Vietnam shows how this force can be relatively weak even in such a large military effort, since the operational approach of sequential large-scale search-and-destroy tactics did not require additional security forces; pacification was deemed as a job for ARVN for most of the war.

Artillery Unit Conversion Cost

The fourth factor which influences the employment of artillery units is the relative cost of converting them from delivering indirect fires to another counterinsurgent role.[7] This factor does not always exhibit a major influence since it only appears if they actually operate in a nonstandard role. Nonetheless, it is a major consideration with potential long-term effects, and therefore worthy of inclusion with the other three factors. In OIF, both the short-term effect in Iraq and the long-term implications of this cost were evident. In Malaya and Vietnam these costs were present, but very manageable. Artillery units only converted to the extent of enabling indirect fires with theater-specific requirements for dispersion and coverage, requirements for which they were not initially prepared to satisfy.

The Fundamentals of Employing Artillery Units in Counterinsurgency

The preceding analyses of artillery units in three counterinsurgencies indicate that there are several recurring best practices or deficiencies in the effort. As with the factors which affect their employment, these trends are broadly descriptive themes and not prescriptive principles. These factors represent the environment in which commanders must decide to employ artillery units, while the fundamentals represent the criteria and requisites to implement that decision. Those fundamentals are best described in five distinct actions:

1. Invest in artillery units' tactical leadership.
2. Exploit lessons learned.
3. Support the operational approach and strategic framework.
4. Maintain a pragmatic fire support capability.
5. Minimize collateral damage.

These are not solely applicable to counterinsurgent warfare; they also support the broader imperatives of artillery integration in combined arms maneuver and wide area security to seize, retain, and exploit the initiative in all unified land operations. Additionally, they are not simply actions to be taken during a counterinsurgency; they do not require a change in mindset as operations begin. The following actions must pervade training and preparations prior to a counterinsurgency operation in order to be truly effective.

Invest in Artillery Units' Tactical Leadership

Decentralized operations and the imperatives of a harmonized effort in the counterinsurgency environment demand junior leaders who can analyze a situation, gain critical understanding, and execute a feasible solution. The challenge is even higher in the artillery, since leaders must combine these abstract cognitive qualities with the tactical and technical competence to deliver and integrate accurate predicted fires. Tactical leadership provided artillery units in Malaya and Vietnam with the ability to operate from distributed locations. In Malaya, leaders were able to overcome the tyranny of distance between units and establish firing capabilities in elements as small as two guns. Leaders in Vietnam were able to develop new techniques to enable firing operations from the FSBs and innovative methods to defend them. But the effect of leaders in artillery units may have its greatest illustration in OIF. Leaders were not only able to maintain firing capabilities when needed, but also provided the backbone of the organization as they switched roles. Effective tactical leaders provide this inherent flexibility.

Exploit Lessons Learned

Artillery units must be able to exploit the lessons learned from previous units in counterinsurgencies. This ability is critical since they may be employed in a role it has no organizational experience with, or at least in an environment which seems very foreign and restrictive when compared to a linear battlefield. At a minimum, familiarity with lessons learned in combat provide artillery units with an improved appreciation of the demands in counterinsurgency warfare. This is a glaring deficiency with both OIF and the Malayan Emergency.[8] As discussed in Chapter Five, there is still a paucity of lessons exploited in any mode of artillery unit employment. At the very least, they should be accessible. The Malayan Emergency provides a good example of what can happen when lessons learned are institutionally compartmentalized. The lessons were exploited within the Far East theater through an institutional and iterative process, but were not shared with the wider British Army. As a result, few British officers (even gunners) are even aware that the guns were used extensively in places such as Malaya and Dhofar as a part of the counterinsurgency effort.[9]

Support the Operational Approach and Strategic Framework

The employment of artillery units must be a derivative of the operational approach to counterinsurgency. In counterinsurgencies, discrete actions may have unintended strategic effects, so their employment must also be sensitive to the overall strategic context. The implication of this fundamental is that an ill-suited operational approach may very well use them in a role which deepens its negative effects. In that case the entire operational approach must be revised, not just a component of it such as the artillery. This was the case in the early stages of both Malaya and Vietnam, when artillery units supported large-scale search-and-destroy missions. When the operational approach in Malaya changed to isolating the CTs from their base of support in the Min Yuen at the village level, the role of the artillery also changed and contributed to the success of that structured approach. Vietnam exemplifies another facet of this fundamental. Although employing artillery units as pacification-focused task forces may have worked well to address the root of the popular grievances and rural security in SVN, American infantry was so decentralized that they were not postured to succeed against VC main forces and NVA units without the coverage and responsiveness of artillery's indirect fires. OIF provides yet another example, since a gradual change to the operational approach of securing the population required a change in the general employment of direct support artillery units, either by split operations or a complete conversion to a nonstandard security role.

Maintain a Pragmatic Fire Support Capability

Artillery units must support a flexible fire support network that works within the counterinsurgency environment with a level of firepower commensurate to meet the insurgent threat. This does not necessarily mean that they have to be a part of this network, if their core competencies of timeliness, accuracy, and reliability are ensured by a professional artillerymen's ability to integrate other sources of firepower. This will generally manifest itself as a mix of both direct support and general support fires, as well as a mix of land-based and aerial firepower. This fundamental is closely linked to the factor of counterinsurgent force organization, since a sufficient fire support capability will allow some artillery units to be employed for other means. This may also require extremely adaptive techniques to support maneuver forces with firepower in a counterinsurgency, lending further credence to the notion that "relationships between commanders are more important than command relationships."[10]

The Malayan Emergency exhibits a unique application of this fundamental. Although decentralized batteries could not cover all operations with supporting fires, this was not necessarily required. Mortars were sufficient for most small patrols, so artillery units were used to sequentially support prioritized missions. In Vietnam, an incredible amount of firing elements were required to establish a sufficient fire support network against an enemy that periodically massed to conduct large-scale conventional attacks from a position of relative advantage.

Minimize Collateral Damage

The requirement to limit collateral damage in a counterinsurgency is not a new concept, nor is it specific to indirect firepower. However, it is a fundamental of employing artillery units since a population's cultural, social, and economic sensitivities to collateral damage must be determined before employing them to provide massive amounts of indirect fires. This is not to say that artillery fires inherently cause collateral damage. Indirect fires are only one part of a holistic effort in counterinsurgency to minimize collateral damage, but artillery units must take the lead in its reduction. Within the counterinsurgency environment, PGMs are the means to that end, but not an end in and of themselves. Counterinsurgents must concurrently mitigate target location error, insufficient coordination measures, and uninformed decision-making when electing to use any form of firepower among the population.

This fundamental is easily identifiable in most case studies, since the second- and third-order effects of collateral damage are usually noteworthy. In Vietnam, collateral damage was avoided but generally acknowledged as an acceptable cost to the American military, as evidenced in the high rate of unobserved H&I fires. In Malaya, the anecdote of the "Yellow Pin" in Chapter Three illustrates one commander's attempt to isolate target location errors and improve coordination with aerial assets since he sought to leverage its firepower, but was not willing to risk unpredictable strikes. OIF shows a mature implementation of this fundamental, as American forces adapted to mitigate collateral damage through CDE methods and precision munitions to include GMLRS and Excalibur.

Conclusion: The Effect of Educated Leaders in Counterinsurgency

Leaders impact all five fundamentals for employing artillery units within a counterinsurgency effort. Leaders' contribution to the first two fundamentals is that they can imbue a unit with experience, vision, and flexibility. To ensure that they are supporting a sound operational approach, leaders use their understanding of commander's intent and empathetic feedback from their environment. The relationships between leaders and their ability to adapt organizational models has the greatest impact on establishing a pragmatic fire support system. Finally, artillery leaders' attention to detail and interpersonal abilities as fire supporters are a key to improving the precision of fires and limiting collateral damage.

With the recognition that effective tactical leadership is an avenue to improve the artillery's capabilities in these fundamental areas, it should be the priority of resources.[11] Training artillery units to deliver timely and accurate indirect fires is an integral part of this effort and not a parallel track, since junior leadership is intimately involved in the art and science of fire support, fire direction, and firing unit operations. But there is another element to waging war in a counterinsurgency that demands further education for leaders, not just positivistic training at centralized professional courses. Due to technological and societal pressures, operational and strategic contexts did not simply change on the margins in the twentieth century, they changed fundamentally.[12] While the *nature* of war itself is unchanging, the *conduct* of warfare changes as a result of these influences. Since any form of warfare is likely to be fought among modern populations in this increasingly complex environment, modern warfare demands more than just a marginal change to the institutional approach for leader education. It requires a fundamental change.

However, most modern armies seek to improve their junior leaders with discrete efforts such as foreign language training to increase cultural empathy, or individual classes which expose leaders to the aspects of civil governance.[13] Even the baseline models for leader training are lacking, as illustrated in a 2010 study which found that only a third of the US Army's Captain's Career Courses achieved academic excellence.[14] A fundamental shift in leader development from training to education is in order, one that rewards efforts to improve critical thinking abilities and cognitive development through a variety of academic endeavors or broadening experiences. Education "allows one to gain better understanding of experiences and training," so parallel efforts must be made to increase a leader's experience base if they have not ventured from the friendly confines of their home station or an air-conditioned headquarters in Iraq.[15] Leaders in counterinsurgencies cannot metaphorically "think outside of the box" any more than they can kill insurgents by telekinesis. Exhortations to this effect deny the limits of leaders as humans. But an educated tactical leader with a broad base of both education and experience has a "larger box" in which to find his solutions, and will be able to harness more approaches to a problem due to an increased capacity for problem-solving. Instead of trying to educate well-trained leaders with constrained resources, an army should attempt to train well-educated leaders with constrained resources.

Without educated leaders, there is a diminished capacity for creative solutions. If leaders are only trained in distinct tasks, then an army will need prescriptive principles for complex environments like counterinsurgencies, which this study attempts to avoid. These five descriptive fundamentals should serve as guideposts to prepare artillery units for operations in counterinsurgency operations, but in no way should they obviate technical and tactical training. Delivering indirect fires has always been a technically and tactically demanding endeavor, and it will remain that way for the foreseeable future. Artillery leaders must understand the sciences of manual gunnery and terminal effects. As with an effort to educate leaders, exercising and developing artillery units along the lines of these five fundamentals will benefit high-intensity conventional capabilities as well as counterinsurgency capabilities.

A temptation is to contend that OIF illustrates a situation where artillery units must be prepared to conduct a variety of missions in counterinsurgencies, and at the very least they must adapt to unforecasted requirements for delivering indirect fires. But the impact is much more profound; the entire history of artillery units in counterinsurgencies bears this out, with OIF serving as just the most recent and vivid example. As General Raymond Odierno noted during a 2004 interview with *Field Artillery*, "artillery has to be a versatile asset. The Army can no longer afford to have artillerymen just do artillery missions. So Redlegs also must be able to set up flash checkpoints, patrol, conduct cordon and search operations, etc."[16] Accordingly, the Army must maintain the big stick of artillery as a flexible option for their next unforecasted and amorphous conflict.

Notes

1. Chapter Two discussed the requirement for indirect fires in depth. It arises from the dispersion of insurgents and the frequency of contact between them and security forces.

2. Janice E. McKenney, *The Organizational History of Field Artillery 1775–2003.* (Washington, DC: US Army Center for Military History, 2007), 269; John J. McGrath, *Fire for Effect: Field Artillery and Close Air Support in the US Army* (Fort Leavenworth, KS: Combat Studies Institute Press, 2007), 117. The overall force consisted of 61 artillery battalions and only 59 infantry battalions.

3. In accordance with the discussion in Chapter Two, constraints are the restrictions that a higher echelon places on a lower echelon, whether political or military; limitations are physical restrictions on the application of indirect fires.

4. Chapter Two's description of this factor discusses the "pushing" and "pulling" forces in depth.

5. While no statistical breakdown of counterinsurgent force composition exists to support this without further research, this is the author's qualitative assessment considering the following counterinsurgency campaigns: American forces in The Phillippines, British forces in Ireland, British forces in the Malayan Emergency, French forces in Algeria, American and GVN forces in Vietnam, British and Omani forces in Dhofar, Rhodesian forces in the Rhodesian Bush Wars, British forces in Northern Ireland, Coalition forces in Iraq (OIF), and Coalition forces in Afghanistan (OEF).

6. Chapter Three discusses that the "push" force from improved mortars was arguably more important than air assets' contributions.

7. Chapter Two discusses this factor, and the fact that the cost is expressed in many terms such as time, physical resources, and the opportunity cost of degrading indirect fires capabilities over the long-term.

8. The presentation for the 2011 Fires Seminar "State of Fires" briefing cites combat leader debriefings as an input to doctrinal reform, not as a source in and of itself. US Army Fires Center of Excellence, *2011 Fire Seminar: State of Fires* (Symposium at Cameron University, Lawton, OK, 17 May 2011), slide 12.

9. This is based on the author's discussions with British Army officers in the UK, March and April 2011.

10. AA402, Special Forces Group Commander, Interview by Brian McCarthy and Jesse Stewart, Fort Lewis, WA, 31 August 2010.

11. The 2011 Fires Seminar "State of Fires" briefing concludes that solving the current gaps in capabilities is "not just about material." Although leader education is not specifically addressed, this notion is a step in the right direction for resource allocations to improve artillery capabilities. US Army Fires Center of Excellence, *2011 Fire Seminar: State of Fires* (Symposium at Cameron University, Lawton, OK, 17 May 2011), slide 14.

12. Author Fathali Moghaddam provides a compelling illustration of this shift in *The New Global Insecurity.* He traces the fundamental shift in insecurity in the world to the force of fragmented globalization, which he studies through

the components of resource insecurity and religious insecurity. Fathali M. Moghaddam, *The New Global Insecurity* (Santa Barbara, CA: Praeger Security International, 2010), 19-20.

13. Based on the author's experience with the American army, interviews with British army officers in their institutions of higher learning in 2011, and individual discussions on the subject with Australian, Canadian, and French officers in 2010 and 2011.

14. William M. Raymond, Keith R. Beurskens, and Steven M. Carmichael, "The Criticality of Captains' Education, Now and in the Future," *Military Review* (November-December 2010): 51.

15. This quote is taken from William M. Raymond et al., "The Criticality of Captains' Education, Now and in the Future," *Military Review* (November-December 2010): 56. As a point of fact, the author enjoys his air conditioning on deployments to the Middle East. But to address the matter directly; in any large war effort there will be some officers who simply do not have the same opportunities as their peers to experience conventional or counterinsurgent warfare since there is a valid requirement to man headquarters echelons, institutional army posts, and contingency forces. Unfortunately, there is no provision for the secondment of officers, which directly assists an allied security force while simultaneously providing an officer with invaluable experience.

16. Patricia Slayden Hollis, "Division Operations Across the Spectrum-Combat to SOSO in Iraq: Interview with Major General Raymond T. Odierno, CG of 4th ID in OIF," *Field Artillery* (March-June 2004): 11.

Bibliography

Primary Sources

Interviews

Command and General Staff College (CGSC) Scholars Program 2011. Scholars Program *Counterinsurgency Research Study 2011*. Research Study, Fort Leavenworth, KS: Ike Skelton Chair in Counterinsurgency, 2011. This study included interviews of counterinsurgency practitioners and policy professionals from the United States and United Kingdom. Each interview was executed as an oral history interview and adhered to Army policies of informed consent in compliance with federal law. Finally, each interview was coordinated through the Ike Skelton Chair in Counterinsurgency, CGSC Fort Leavenworth, KS.

Boston, Massachusetts

BF010, Former Army Officer. Interview by Richard Johnson and Aaron Kaufman, 11 March 2011.
BF020, Civilian Adviser to MNF-I. Interview by Richard Johnson and Aaron Kaufman, 11 March 2011.
BF030, Battery Commander. Interview by Richard Johnson and Aaron Kaufman, 12 March 2011.
BF040, Battery Commander. Interview by Richard Johnson and Aaron Kaufman, 14 March 2011.

Fort Bliss, Texas

BB010, Battalion Commander. Interview by Mark Battjes and Nathan Springer, 2 March 2011.
BB020, Battalion Commander. Interview by Mark Battjes and Nathan Springer, 2 March 2011.
BB030, Brigade Commander. Interview by Mark Battjes and Nathan Springer, 3 March 2011.

Fort Bragg, North Carolina

AA610, Platoon Leader and FDO. Interview by Travis Molliere and Carrie Przelski, 18 August 2010
BC010, Field Grade Officer. Interview by Robert Green and Aaron Kaufman, 1 March 2011.
BC020, Brigade Commander. Interview by Robert Green and Aaron Kaufman, 2 March 2011.
BC030, Battalion Commander. Interview by Benjamin Boardman and Richard Johnson, 1 March 2011.
BC040, Battalion Commander. Interview by Benjamin Boardman and Richard Johnson, 2 March 2011.
BC050, Battalion Commander. Interview by Benjamin Boardman and Richard Johnson, 2 March 2011.
BC060, Battalion Commander. Interview by Benjamin Boardman and Richard Johnson, 3 March 2011.

Fort Irwin, California

BE010, Transition Team Leader. Interview by Mark Battjes and Thomas Walton, 7 March 2011.
BE020, Transition Team Member. Interview by Mark Battjes and Thomas Walton, 7 March 2011.
BE030, Company Commander. Interview by Mark Battjes and Thomas Walton, 8 March 2011.
BE040, Transition Team Leader. Interview by Mark Battjes and Thomas Walton, 9 March 2011.
BE050, Battery Commander. Interview by Robert Green and Aaron Kaufman, 8 March 2011.
BE060, Brigade Commander. Interview by Mark Battjes and Thomas Walton, 9 March 2011.
BE070, Field Grade Officer. Interview by Robert Green and Aaron Kaufman, 9 March 2011.
BE080, Battalion Commander. Interview by Robert Green and Aaron Kaufman, 7 March 2011.
BE090, Battalion Commander. Interview by Robert Green and Aaron Kaufman, 7 March 2011.

Fort Knox, Kentucky

BD010, Field Grade Officer. Interview by Benjamin Boardman and Dustin Mitchell, 14 March 2011.
BD020, Commander. Interview by Benjamin Boardman and Dustin Mitchell, 14 March 2011.
BD030, Commander. Interview by Benjamin Boardman and Dustin Mitchell, 14 March 2011.
BD040, Commander. Interview by Benjamin Boardman and Dustin Mitchell, 15 March 2011.
BD050, Commander. Interview by Benjamin Boardman and Dustin Mitchell, 15 March 2011.
BD060, Field Grade Officer. Interview by Benjamin Boardman and Dustin Mitchell, 16 March 2011.
BD070, Field Grade Officer. Interview by Benjamin Boardman and Dustin Mitchell, 16 March 2011.
BD080, Field Grade Officer. Interview by Benjamin Boardman and Dustin Mitchell, 17 March 2011.

Fort Leavenworth, Kansas

AA509, Division Commander. Interview by Ken Gleiman and Karsten Haake, 25 August 2010.
AA510, Former DivArty Commander. Interview by Travis Molliere and Jesse Stewart, 4 November 2010.
BA010, Brigade Commander. Interview by Richard Johnson and Thomas Walton, 22 February 2011.
BA020, Battalion Commander. Interview by Mark Battjes and Benjamin Boardman, 23 February 2011.
BA030, Vietnam Veteran. Interview by Aaron Kaufman and Dustin Mitchell, 24 February 2011.
BA040, Brigade Commander. Interview by Aaron Kaufman and Dustin

Mitchell, 23 February, 2011.

BA050, Battalion Commander. Interview by Robert Green and Nathan Springer, 23 February, 2011.

BA060, Battalion Commander. Interview by Robert Green and Nathan Springer, 23 February, 2011.

BA070, Battery Commander. Interview by Richard Johnson and Thomas Walton, 24 February 2011.

BA080, Counterinsurgency Adviser. Interview by Richard Johnson and Nathan Springer, 9 March 2011.

BA090, Brigade Commander. Interview by Mark Battjes and Benjamin Boardman, 24 February 2011.

Afghanistan Panel Discussion, CGSC Art of War Scholars Group and OEF Veterans, 4 February 2011.

Algeria Panel Discussion, CGSC Art of War Scholars Group, 11 January 2011.

Counterinsurgency I Panel Discussion, CGSC Art of War Scholars Group, 30 November 2010.

Counterinsurgency II Panel Discussion, CGSC Art of War Scholars Group, 3 December 2010.

Dhofar Panel Discussion, CGSC Art of War Scholars Group and Dhofar Campaign Veteran, 21 January 2011.

Iraq Panel Discussion, CGSC Art of War Scholars Group and OIF Veterans, 3 February 2011.

Ireland Panel Discussion, CGSC Art of War Scholars Group, 17 December 2010.

Malaya I Panel Discussion, CGSC Art of War Scholars Group, 4 January 2011.

Malaya II Panel Discussion, CGSC Art of War Scholars Group, 7 January 2011.

Northern Ireland Panel Discussion, CGSC Art of War Scholars Group and Operation Banner Veteran, 28 January 2011.

Philippines Panel Discussion, CGSC Art of War Scholars Group, 6 December 2010.

Rhodesia Panel Discussion, CGSC Art of War Scholars Group and Rhodesian Bush War Veteran, 25 January 2011.

Vietnam I Panel Discussion, CGSC Art of War Scholars Group and Vietnam War Veteran, 14 January 2011.

Vietnam II Panel Discussion, CGSC Art of War Scholars Group and Vietnam War Veteran, 18 January 2011.

Fort Lewis, WA

AA402, Special Forces Group Commander. Interview by Brian McCarthy and Jesse Stewart, 31 August 2010. Fort Lewis, WA.

Fort Stewart, Georgia

BG020, Brigade Commander. Interview by Mark Battjes and Thomas Walton, 14 March 2011.

BG030, Troop Commander. Interview by Mark Battjes, Nathan Springer, and Thomas Walton, 14 March 2011.

BG040, Brigade Commander. Interview by Nathan Springer and Thomas Walton, 15 March 2011.

BG050, Battalion Commander. Interview by Mark Battjes, 15 March 2011.

BG060, Battalion Commander. Interview by Mark Battjes, 15 March 2011.

BG070, Field Grade Officer. Interview by Nathan Springer and Thomas Walton, 15 March 2011.

BG080, Battalion Commander. Interview by Mark Battjes and Thomas Walton, 16 March 2011.

BG090, Battalion Commander. Interview by Mark Battjes and Nathan Springer, 16 March 2011.

BG100, Brigade Commander. Interview by Mark Battjes and Nathan Springer, 16 March 2011.

United Kingdom

AA1009, Retired General Officer. Interview by Ken Gleiman, Brian McCarthy, Travis Molliere, and Carrie Przelski, 4 October 2010. United Kingdom.

AA1012, British General Officer. Interview by Mike Dinesman, Ken Gleiman, Brian McCarthy, Jesse Stewart, 22 September 2010. United Kingdom

AA1013, British General Officer. Interview by Karsten Haake, Matt Marbella, Travis Molliere, and Carrie Przelski, 23 September 2010. United Kingdom

BI010, Senior British Officer. Interview by Mark Battjes, Benjamin Boardman, Robert Green, Richard Johnson, Aaron Kaufman, Dustin Mitchell, and Nathan Springer, 29 March 2011.

BI020, Battle Group Commander. Interview by Aaron Kaufman and Thomas Walton, 31 March 2011.

BI030, Field Grade Officer. Interview by Robert Green and Thomas Walton, 29 March 2011.

BI040, Field Grade Officer. Interview by Mark Battjes and Dustin Mitchell, 1 April 2011.

BI050, Dhofar Veterans Panel. Interview by Mark Battjes, Ben Boardman, Robert Green, Richard Johnson, Aaron Kaufman, Dustin Mitchell, Nathan Springer, and Thomas Walton, 28 March 2011.

BI060, Dhofar Veterans Panel. Interview by Mark Battjes, Ben Boardman, Robert Green, Richard Johnson, Aaron Kaufman, Dustin Mitchell, Nathan Springer, and Thomas Walton, 2 April 2011.

BI070, Retired General Officer. Interview by Interview by Mark Battjes, Ben Boardman, Robert Green, Richard Johnson, Aaron Kaufman, Dustin Mitchell, Nathan Springer, and Thomas Walton, 30 March 2011.

BI080, Retired General Officer. Interview by Ben Boardman, Robert Green, Nathan Springer, and Thomas Walton, 3 April 2011.

BI090, Retired General Officer. Interview by Ben Boardman, Robert Green, Nathan Springer, and Thomas Walton, 4 April 2011.

BI100, Senior Army Officer. Interview by Mark Battjes, Richard Johnson, Aaron Kaufman, and Dustin Mitchell, 4 April 2011.

BI110, Battalion Commander. Interview by Mark Battjes, Richard Johnson, and Dustin Mitchell, 8 April 2011.

BI120, Retired Army Officer. Interview by Ben Boardman, Robert Green, Nathan Springer, and Thomas Walton, 8 April 2011.

BI130, Platoon Commander. Interview by Ben Boardman and Richard Johnson, 5 April 2011.

BI140, Afghan Army Adviser. Interview by Ben Boardman and Richard Johnson, 5 April 2011.

BI150, Company Sergeant Major. Interview by Aaron Kaufman and Dustin Mitchell, 5 April 2011.
BI160, Company 2nd In Command. Interview by Aaron Kaufman and Dustin Mitchell, 5 April 2011.
BI170, Afghan Army Adviser. Interview by Aaron Kaufman and Dustin Mitchell, 5 April 2011.
BI190, Senior Non-Commissioned Officer. Interview by Mark Battjes and Thomas Walton, 5 April 2011.
BI200, Platoon Commander. Interview by Aaron Kaufman and Dustin Mitchell, 7 April 2011.
BI210, Company 2nd In Command. Interview by Mark Battjes and Thomas Walton, 7 April 2011.
BI220, Field Grade Officer. Interview by Aaron Kaufman and Dustin Mitchell, 7 April 2011.
BI230, Company Commander. Interview by Robert Green and Nathan Springer, 7 April 2011.
BI240, Company Grade Officer. Interview by Benjamin Boardman and Richard Johnson, 7 April 2011.
BI250, Battalion Commander. Interview by Benjamin Boardman and Richard Johnson, 7 April 2011.
BI260, Non-Commissioned Officer. Interview by Robert Green and Nathan Springer, 7 April 2011.
BI270, Company Grade Officer. Interview by Mark Battjes and Thomas Walton, 7 April 2011.
BI280, Commander's Panel. Interview by Richard Johnson, 1 April 2011.
BI290, Battery Commander. Interview by Richard Johnson, 1 April 2011.
BI300, Company Commander. Interview by Richard Johnson, 2 April 2011.
BI310, Company Commander. Interview by Benjamin Boardman and Nathan Springer, 31 March 2011.
BI320, Field Grade Officer. Interview by Benjamin Boardman and Dustin Mitchell, 29 March 2011.
BI330, Dhofar Veteran. Interview by Robert Green, 28 March 2011.

Washington, DC

AA810, Battalion Commander. Interview by Ken Gleiman, Matt Marbella, Brian McCarthy, and Travis Molliere, 13 September 2010. Washington, DC.
BH010, Senior Policy Official. Interview by Mark Battjes, Ben Boardman, Robert Green, Richard Johnson, Aaron Kaufman, Dustin Mitchell, Nathan Springer, and Thomas Walton, 21 March 2011.
BH020, Field Grade Officer. Interview by Mark Battjes, Ben Boardman, Robert Green, Richard Johnson, Aaron Kaufman, Dustin Mitchell, Nathan Springer, and Thomas Walton, 21 March 2011.
BH030, Iraq Veterans Panel. Interview by Mark Battjes, Robert Green, Aaron Kaufman, and Dustin Mitchell, 22 March 2011.
BH040, Afghanistan Veterans Panel. Interview by Richard Johnson, Aaron Kaufman, Nathan Springer, and Thomas Walton, 24 March 2011.
BH050, Historian. Interview by Mark Battjes, Robert Green, Richard Johnson, Aaron Kaufman, and Dustin Mitchell, 22 March 2011.
BH060, Vietnam Political and Military Analyst. Interview by Mark Battjes, Ben

Boardman, Robert Green, and Dustin Mitchell, 22 March 2011.
BH070, Iraqi Mayor. Interview by Mark Battjes and Robert Green, 25 March 2011.

Other Oral Histories

Garcia, Lynne. *Interview with Colonel Frank Hull.* Fort Leavenworth, KS: Combat Studies Institute, 2005.

Hollis, Patricia Slayden. "Division Operations Across the Spectrum - Combat to SOSO in Iraq: Interview with Major General Raymond T. Odierno, CG of 4th ID in OIF." *Field Artillery* (March-June 2004): 10-14.

Official Reports and Memoranda

1st Infantry Division. *After Action Report-Operation Junction City.* 1967.

4th Infantry Division Artillery. *Operational Report - Lessons Learned, Headquarters 4th Infantry Division Artillery, Period Ending 31 Jan 1970.* 1970.

25th Infantry Division Artillery. *Operational Report of 25th Inf Div Arty for Period Ending 31 Jul 68.* 1968.

101st Airborne Division (Airmobile). *Combat Operations After Action Report, Operation Randolph Glen.* 1970.

———. *Operational Report-Lessons Learned, 101st Airborne Division (Airmobile), Period Ending 31 October 1970.* 1970.

173rd Airborne Brigade. *Combat After Action Report-Operation Niagara / Cedar Falls, Period 5-26 Jan 1967.* 1967.

———. *Combat After Action Report-Operation Junction City I.* 1967.

Baker-Hamilton Commission. *Iraq Study Group Report: Gravel Edition.* Washington, DC: Filiquarian Publishing, 2006.

Cushman, John H. *Senior Officer Debriefing Report: Major General John H. Cushman, Commander, Delta Regional Assistance Command, Vietnam, 14 May 71-14 Jan 72.* Washington, DC: Department of the Army, 1972.

Deputy Chief of Staff for Military Operations - U.S. Department of Defense. *A Program for the Pacification and Long-Term Development of South Vietnam,* 2 vol. Washington, DC: Department of Defense, 1966.

District Office of Kuala Lipis. *Lipis DWEC Operation No. 2: Operation Apollo.* 1954.

Forster, J. M. *A Comparative Study of the Emergencies in Kenya and Malaya.* Operational Research Unit Far East, 1957.

Gates, Robert M. *Request to Change the Name of Operation IRAQI FREEDOM to Operation NEW DAWN.* 2010.

Hoang Ngoc Lung. *Indochina Monographs: Strategy and Tactics.* Washington, DC: US Army Center for Military History, 1978.

Hosmer, Stephen T., and Sibylle O. Crane. "Counterinsurgency, A Symposium 16-20 April 1962." RAND Corporation Reports, Washington, DC, 1962.

Kagan, Frederick W. *Choosing Victory: A Plan for Success in Iraq.* Washington, DC: American Enterprise Institute, 2006.

MacFarland, Sean, Michael Shields, and Jeffrey Snow. The King and I: The Impending Crisis in the Field Artillery's Ability to Provide Fire Support to Maneuver Commanders. Unpublished.

McWilliams, Timothy S. *Al-Anbar Awakening: U.S. Marines and Counterinsurgency in Iraq, 2004-2009*. Quantico, VA: Marine Corps University Press, 2009.
Military Assistance Command, Vietnam, United States, *Command History 1968 Volume I*. Saigon: 1969.
———. *Lessons Learned Number 31: Artillery Organization and Employment in Counter Insurgency.* Saigon: 1963.
Ngo Quang Truong. *Indochina Monographs: Territorial Forces.* Washington, DC: Center of Military History, 1981.
State Secretariat of Central Perak. *Central Perak SWEC Operation Order No. 1-57: Operation 'Ginger.'* 1957.
State Secretariat of Pahang. *Pahang SWEC Operation No. 2: Operation Apollo*. 1954.
Sunderland, Riley. *Army Operations in Malaya, 1947-1960*. Santa Monica, CA: RAND, 1964.
———. *Organizing Counterinsurgency in Malaya, 1947-1960*. Santa Monica, CA: RAND, 1964.
———. *Winning the Hearts and Minds of the People: Malaya 1948-1960*. Santa Monica, CA: RAND, 1964.
Tran Dinh Tho. *Indochina Monographs: Pacification*. Washington, DC: Center for Military History, 1980.

Personal Accounts

Baker, Jay B. "Tal Afar 2005: Laying the Counterinsurgency Framework." *Army* (June 2009): 61-67.
Bidwell, R. G. S. "Gunner Tasks in Southeast Asia." *Royal Journal of Artillery* 91, no.2 (September 1964): 85-96.
Beeton, C. "Locating in Borneo." *Royal Journal of Artillery* 96, no. 1 (March 1969): 14-18.
Bremer, L. Paul. *My Year in Iraq*. New York: Simon and Schuster, 2006.
Chiarelli, Peter, and Patrick Michaelis. "The Requirements for Full-Spectrum Operations." *Military Review* (July-August 2005): 4-17.
Clutterbuck, Richard L. *The Long, Long War.* New York: Praeger, 1966.
Colby, William. *Honorable Men: My Life in the CIA*. New York: Simon and Schuster, 1978.
Collier, Craig. "Now that we're leaving Iraq, what did we learn." *Military Review* (September-October 2010): 88-93.
Corum, James. *Training Indigenous Forces in Counterinsurgency: A Tale of Two Insurgencies*. Carlisle, PA: Strategic Studies Institute, 2006.
Daniel, T. W. "Man Hunting in Malaya." *The Royal Journal of Artillery* 77, no. 3 (July 1950): 257-260.
Ferry, J. P. "Full Employment." *Royal Journal of Artillery* 92, no 1. (March 1965): 21-29.
Franks, Tommy. *American Soldier.* New York: Regan Books, 2004.
Giap, Vo Nguyen. *How We Won The War.* Philadelphia, PA: RECON Publications, 1976.
———.*Inside the Vietminh: Vo Nguyen Giap on Guerrilla War*. Quantico, VA: Marine Corps Association, 1962.
———. *The Military Art of People's War*, edited by Russell Stetler. New York:

Monthly Review Press, 1970.
Guevara, Ernesto "Che." *Guerrilla Warfare*, edited by Brian Loveman and Thomas M. Davies. Lincoln, NE: University of Nebraska Press, 1985.
Haldane, James Aylmer. *The Insurrection in Mesopotamia*. London: The Imperial War Museum, 2005.
Hennicker, M. C. A. *Red Shadow Over Malaya*. London: William Blackwood and Sons, 1955.
Hunt, Ira A. *The 9th Infantry Division in Vietnam: Unparalleled and Unequaled*. Lexington, KY: University of Kentucky Press, 2010.
Jeapes, Tony. *SAS Secret War: Operation Storm in the Middle East*. London: Greenhill Books, 2005.
Komer, Robert. *Bureaucracy at War: U.S. Performance in the Vietnam Conflict*. Boulder, CO: Westview Press, 1986.
Lamb, James Daniel. "25th Field Regiment in Malaya." www.britains-smallwars.com/ malaya/malaya-25field%20reg/25_fieldregt.html (accessed 6 June 2011).
Lansdale, Edward. "Contradictions in Military Culture." In *The Lessons of Vietnam*, edited by W. Scott Thompson and Donaldson Frizzel, 40-46. New York: Crane, Russak and Company, 1977.
Lindsay, Patrick J. "Counterguerrilla Warfare." *Artillery Trends* 29 (February 1964): 11-24.
Mansoor, Peter R. *Baghdad at Sunrise: A Brigade Commander's War in Iraq*. New Haven, CT: Yale University Press, 2008.
Miers, Richard. *Shoot To Kill*. London: Faber and Faber, 1959.
Mitchell, Joshua D. "Afghanistan: Firing Artillery Accurately with Air Force Met Support." *Field Artillery* (January-February 2003): 38-41.
Rowe, G.C.K. "Presentation of a Silver Centre-piece to the Federation Artillery." *The Royal Journal of Artillery* 87, no. 3 (Summer 1960): 128-129.
Sanchez, Ricardo. *Wiser in Battle*. New York: HarperCollins, 2008.
Sky, Emma. "Iraq, From Surge to Sovereignty." *Foreign Affairs* (March-April 2011). www.foreignaffairs.com/articles/67481/emma-sky/iraq-from-surge-to-sovereignty (accessed 20 May 2011).
Sorley, Lewis. *Vietnam Chronicles: The Abrams Tapes 1968-1972*. Lubbock, TX: Texas Tech University Press, 2004.
Smith, Neil, and Sean MacFarland. "Anbar Awakens: The Tipping Point." *Military Review* (March-April 2008): 41-53.
Sweeney, Patrick J., and Jason C. Montgomery. "Iraq: 101st Division LNO in the V Corp FECC." *Field Artillery* (July-August 2003): 40-44.
Torrance, Thomas G., and Noel T. Nicolle. "Observations From Iraq: The 3rd Div Arty in OIF." *Field Artillery* (July-August 2003): 30-35.
Truong Nhu Tang, *A Vietcong Memoir*. San Diego: Harcourt Brace Jovanovich Publishers, 1985.
The Virgin Soldiers. "British and Commonwealth Units That Served in the Malayan Emergency." www.britains-smallwars.com/malaya/reg.html (accessed 6 June 2011).
Westmoreland, William C. "A Military War of Attrition." In *The Lessons of Vietnam*, edited by W. Scott Thompson and Donaldson Frizzel, 57-71. New York: Crane, Russak and Company, 1977.

———. *A Soldier Reports.* New York: Da Capo Press, 1976.

Wilford, D. J. "Some Aspects of Anti-Terrorist Operations: Malaya." Fort Bragg, NC, 7th Special Forces Group Operational Assessment, 1963.

Doctrinal References

1st Infantry Division. *Soldier's Handbook to Iraq.* Wurzburg, GE: 1st Infantry Division, 2004.

FARELF Training Center, *The Conduct of Anti-Terrorist Operations in Malaya.* Saint Petersburg, FL: Hailer, originally published internally in Malaya in 1958.

Federation of Malaya. *Handbook to Malaya and the Insurgency.* Singapore: Regional Information Office for the United Kingdom in South East Asia, 1953.

Military Assistance Command, Vietnam, United States. *Handbook for Military Support of Pacification.* Saigon: 1968.

———. *PAVN Artillery (Rocket Units)-1967.* Saigon: 1967.

———. *RF–PF Handbook for Advisors.* Saigon: 1969.

U.S. Army Training and Doctrine Command. TRADOC Pamphlet 525-3-4, *U.S. Army Functional Concept For Fires 2016-2028.* Washington, DC: Department of the Army, 2010.

U.S. Department of the Army. Field Manual 1-02, *Operational Terms and Graphics.* Washington, DC: Department of the Army, 2004.

———. Field Manual 3-0, *Operations.* Washington, DC: Department of the Army, 2008.

———. Field Manual 3-09.21, *Tactics, Techniques and Procedures for the Field Artillery Battalion.* Washington, DC: Department of the Army, 2001.

———. Field Manual 3-24, *Counterinsurgency.* Washington, DC: Department of the Army, 2006.

———. Field Manual 3-24.2, *Tactics in Counterinsurgency.* Washington, DC: Department of the Army, 2009.

———. Field Manual 31-16, *Counterguerrilla Operations.* Washington, DC: Department of the Army, 1963.

———. Field Manual 31-73, *Adviser Handbook for Counterinsurgency.* Washington, DC: Department of the Army, 1965.

———. Field Manual 6-20-10, *Tactics, Techniques and Procedures for the Targeting Process.* Washington, DC: Department of the Army, 1996.

Secondary Sources

Afsar, Shahid, and Chris Samples. "The Taliban: An Organizational Analysis." *Military Review* (May-June 2008): 58-73.

Andrade, Dale. "Westmoreland Was Right: Learning the Wrong Lessons from the Vietnam War." *Small Wars and Insurgencies* 19, no. 2 (June 2008): 145-181.

Bacevich, Andrew. *The New American Militarism: How Americans are Seduced by War.* Oxford, UK: Oxford University Press, 2005.

Bailey, J. B. A. *Field Artillery and Firepower.* Oxford, UK: Oxford Military Press, 1989.

Barber, Noel. *The War of the Running Dogs: How Malaya Defeated the Communist Guerrillas 1948-1960.* London: Cassell, 1971.

Barry, Robert F. "Why Organic Fires?" *Field Artillery* (March-June 2004): 13-18.

Beckett, Ian. "The British Counter-insurgency Campaign in Dhofar, 1965-1975." In *Counterinsurgency in Modern Warfare*, edited by Daniel Marston and Carter Malkasian, 175-190. Oxford, UK: Osprey Publishing, 2010.

Beckerman, Linda P. *The Non-Linear Dynamics of War*. Science Applications International Corporation, 1999.

Bergerud, Eric. *The Dynamics of Defeat: The Vietnam War in Hau Nghia Province*. Boulder, CO: Westview Press, Inc., 1991.

Birtle, Andrew J. *U.S. Army Counterinsurgency and Contingency Operations Doctrine 1860-1941*. Washington, DC: Center of Military History, United States Army, 2003.

———. *U.S. Army Counterinsurgency and Contingency Operations Doctrine 1942-1976*. Washington, DC: Center of Military History, United States Army, 2006.

Boris, Christopher. "Low Angle Fires for MOUT." *FA Journal* (November-December 2001): 42-45.

Boylan, Kevin M. "The Red Queen's Race: Operation Washington Green and Pacification in Binh Dinh Province, 1969-1970." *The Journal of Military History* 73 (October 2009): 1195-1230.

Brock, John. "We Cannot Take Your Call for Fire Right Now: Does the Global War on Terror Signal the Demise of the Field Artillery?" Monograph, School of Advanced Military Studies, Fort Leavenworth, KS, 2006.

Burton, Brian, and John Nagl. "Learning As We Go: The US Army Adapts to COIN in Iraq, July 2004-December 2006." *Small Wars and Insurgencies* 19, no. 3 (September 2008): 303-327.

Cable, Larry E. *Conflict of Myths: The Development of American Counterinsurgency Doctrine and the Vietnam War*. New York: New York University Press, 1986.

Carland, John M. *U.S. Army in Vietnam: Combat Operations, Stemming the Tide, May 1965 to October 1966*. Washington, DC: US Army Center of Military History, 2000.

Carruthers, Lawrence H. "Characteristics and Capabilities of Enemy Weapons." *Artillery Trends* 45 (September 1970): 11 - 24.

Christie, Michael. "Iraqi Civilian Deaths Drop to Lowest Level of War." *Reuters*, 30 November, 2009. www.reuters.com/article/2009/11/30/us-iraq-toll-idUSTRE5AT3ZE20091130 (accessed 20 May 2011).

Clausewitz, Carl von. *On War*, edited and translated by Michael Howard and Peter Paret. Princeton, NJ: Princeton University Press, 1976.

Coates, John. *Suppressing Insurgency*. Boulder, CO: Westview Press, 1992.

Comber, Leon. *Malaya's Secret Police 1945-1960: The Role of the Special Branch in the Malayan Emergency*. Melbourne, AU: Monash University Press, 2009.

Cordesman, Anthony. *Iraqi Security Forces: A Strategy for Success*. Westport, CT: Praeger Security International, 2006.

Cosgriff, Kenneth, and Richard Johnson, *Organizational Resistance to Change in the Airborne Field Artillery*. Master's thesis, Webster University, 2006.

Cosmas, Graham A. *U.S. Army in Vietnam: MACV, The Joint Command in the*

   Years of Escalation 1962 to 1967. Washington, DC: US Army Center of Military History, 2006.

———. *U.S. Army in Vietnam: MACV: The Joint Command in the Years of Withdrawal 1968-1973.* Washington, DC: US Army Center of Military History, 2005.

Cotton, Sarah K., Ulrich Petersohn, Molly Dunigan, Q Burkhart, Megan Zander-Cotugno, Edward O'Connell, and Michael Webber. *Hired Guns: Views About Armed Contractors in Operation Iraqi Freedom.* Santa Monica, CA: RAND, 2010.

Dehghanpisheh, Babak. "The Great Moqtada Makeover." *Newsweek*, 19 January 2008. www.newsweek.com/2008/01/19/the-great-moqtada-makeover.html (accessed 8 June 2011)

Dahl, John. *Rounders*. Santa Monica, CA: Miramax Films, 1998.

Dastrup, Boyd L. *King of Battle: A Branch History of the U.S. Army's Field Artillery*. Washington, DC: Center of Military History, 1992.

Demarest, Geoff. "Let's Take the French Experience in Algeria Out of U.S. Counterinsurgency Doctrine." *Military Review* (July-August 2010): 19-23.

Dixon, Paul. "Hearts and minds? British counterinsurgency strategy in Northern Ireland." *Journal of Strategic Studies* 32, no. 3 (June 2009): 445-474.

Everett, Patrovik G. "The Role of Field Artillery in Counterinsurgency." Master's Thesis, Command and General Staff College, Fort Leavenworth, KS, 2005.

Fall, Bernard. *Hell in a Very Small Place: The Siege of Dien Bien Phu*. Philadelphia: Lippincott, 1966.

———. *Street Without Joy.* New York: Shocken Books, 1961.

———. *The Two Viet-Nams, A Political and Military Analysis.* New York: Praeger, 1967.

Faugstad, Jesse. "No Simple Solution." *Military Review* (July-August 2010): 32-42.

Foster, Nathaniel w. "Speed Shifting the 155mm Howitzer, Towed: The Evolution of an Idea." *Artillery Trends* 37 (January 1967): 17-22.

Galula, David. *Counterinsurgency Warfare: Theory and Practice.* Saint Petersburg, FL: Glenwood Press, 1964.

———. *Pacification in Algeria, 1956-1958.* Santa Monica, CA: Rand Corporation, 2006.

Gentile, Gian. "A Strategy of Tactics: Population Centric COIN and the Army." *Parameters* (Autumn 2009): 5-17.

Gompert, David C., and John Gordon IV. *War by Other Means*. Santa Monica, CA: RAND Corporation, 2008.

Gordon, Michael R., and Bernard E. Trainor. *Cobra II*. New York: Pantheon Books, 2006.

Gould, Steven J. *Full House: The Spread of Excellence from Plato to Darwin.* New York: Harmony Books, 1996.

Gregorian, Raffi. "Jungle Bashing in Malaya." *Small Wars and Insurgencies* 5, no. 3 (Winter 1994): 338-359.

Green, T. N. *The Guerilla: Selections from the Marine Corps Gazzette.* New York: Praeger, 2005.

Gudmundsson, Bruce I. *On Artillery*. Westport, CT: Praeger, 1993.

Hack, Karl. "The Malayan Emergency as Counter-Insurgency Paradigm." *Journal of Strategic Studies* 32, no. 3 (2009): 383-414.

Hahlweg, Werner. "Clausewitz and Guerrilla Warfare." In *Clausewitz and Modern Strategy*, edited by Michael Handel. London: Frank Cass, 1986, 127-133.

Hart, Peter. *The I.R.A. at War 1916-1923*. Oxford, UK: Oxford University Press, 2003.

Hashim, Ahmed S. "The Insurgency in Iraq." *Small Wars and Insurgencies* 14, no. 3 (August 2003): 3-28.

Heggoy, Alf. *Insurgency and Counterinsurgency in Algeria*. Bloomington, IN: Indiana University Press, 1972.

Heitzke, Kenneth S. "Field Artillery Mission, Weapon, and Organization for Counterguerilla Warfare." Master's Thesis, US Army Command and General Staff College, Fort Leavenworth, KS, 1965.

Herring, George. *America's Longest War*. New York: McGraw-Hill, 1996.

Hoffer, Edward E. "Field Artillery Fire Support for Counterinsurgency Operations: Combat Power or Counterproductive?" Master's Thesis, US Army Command and General Staff College, Fort Leavenworth, KS, 1987.

Hoffman, Bruce. *Lessons for Contemporary Counterinsurgencies: The Rhodesian Experience*. Santa Monica, CA: RAND, 1992.

Hoffman, Frank. "Neo-Classical Counter-insurgency?" *Parameters* (Summer 2007): 71-87.

Hopkinson, Michael. *The Irish War of Independence*. Montreal: McGill - Queen's University Press, 2002.

Horne, Alistair. "The French Army and the Algerian War 1954-62." In *Regular Armies and Insurgency*, edited by Ronald Haycock. London: Croom Helm, 1979, 69-83.

Horwood, Ian. *Interservice Rivalry and Airpower in the Vietnam War*. Fort Leavenworth, KS: Combat Studies Institute Press, 2006.

Howard, Michael. *Clausewitz*. Oxford, UK: Oxford University Press, 1983.

Hunt, Richard. *Pacification: The American Struggle for Vietnam's Hearts and Minds*. Boulder, CO: Westview Press, 1995.

Iron, Richard. "Britain's Longest War: Northern Ireland 1967-2007." In *Counterinsurgency in Modern Warfare*, edited by Daniel Marston and Carter Malkasian, 157-174. Oxford: Osprey Publishing, 2010.

Jackson, Charles W. "Adjust By Sound." *Artillery Trends* 39 (January 1968): 30-34.

Jackson, Joseph A. "Howitzers on the High Ground." Monograph, School of Advanced Military Studies, Fort Leavenworth, KS, 2009.

Jeudwine, Hugh. "A Record of the Rebellion in Ireland in 1920-1, and the Part Played by the Army in Dealing with it (Intelligence)." *British Intelligence in Ireland: The Final Reports*, edited by Peter Hart. Cork: Cork University Press, 2002, 17-60.

Joes, Anthony James. "Counterinsurgency in the Philippines, 1898-1954." In *Counterinsurgency in Modern Warfare*, edited by Daniel Marston and Carter Malkasian, 39-56. Oxford, UK: Osprey Publishing, 2010.

Johnson, Thomas, and Chris M. Maston. "No Sign until the Burst of Fire: Understanding the Pakistan-Afghanistan Frontier." *International*

Security 32, no. 4 (Spring 2008): 41-77.
Jomini, Baron Antoine-Henri de. *Summary of The Art of War, or A New Analytical Compend.* Translated by Winship, O. F. and E. E. McLean. New York: Putnam, 1854.
Khadduri, Majid. *War and Peace in the Law of Islam.* Baltimore, MD: Johns Hopkins Press, 1955.
Kilcullen, David. *The Accidental Guerrilla : Fighting Small Wars in the Midst of a Big One.* New York: Oxford University Press, 2009.
———. *Counterinsurgency.* New York: Oxford University Press, 2010.
Kitson, Frank. *Low Intensity Operations: Subversion, Insurgency, Peacekeeping.* London: Faber and Faber, 1971.
———. *Bunch of Five.* London: Faber and Faber, 1977.
———. *Warfare as a Whole.* London: Faber and Faber, 1987.
Komer, Robert. *The Malayan Emergency in Retrospect: Organization of a Successful Counterinsurgency Effort.* Santa Monica, CA: RAND, 1972.
Krepinevich, Andrew F. Jr. *The Army and Vietnam.* Baltimore, MD: Johns Hopkins University Press, 1986.
Kriger, Norma. *Zimbabwe's Guerrilla War.* Cambridge, UK: Cambridge University Press, 1993.
Ladwig, Walter. "Supporting Allies in COIN: Britain and the Dhofar Rebellion." *Small Wars and Insurgencies* 19, no. 1 (March 2008): 62-88.
Lawrence, T. E. "The 27 Articles of T. E. Lawrence." *The Arab Bulletin* (20 August 1917). www.usma.edu/dmi/IWmsgs/The27ArticlesofT.E.Lawrence.pdf (accessed 8 June 2011).
Levinson, Charles. "An Architect of U.S. Strategy Waits to Pop Cork." *Wall Street Journal*, 27 August 10. online.wsj.com/article/SB10001424052748704125 604575449171253677444.html (accessed 20 May 2011).
Lewis, Bernard. *The Crisis of Islam: Holy War and Unholy Terror.* New York: Random House, 2003.
Linn, Brian. *The Philippine War, 1899-1902.* Lawrence, KS: University of Kansas Press, 2000.
Luce, Don. "Tell Your Friends that We're People." In *The Pentagon Papers Volume V*, edited by Noam Chomsky and Howard Zinn, 91-100. Boston: Beacon Press, 1972.
MacKinlay, John. *The Insurgent Archipelago: From Mao to Bin Laden.* New York: Columbia University Press, 2009.
MacKinlay, John, and Allison Al-Baddawy. *Rethinking Counterinsurgency.* Santa Monica, CA: RAND Corporation, 2008.
Malkasian, Carter. "Counterinsurgency in Iraq." In *Counterinsurgency in Modern Warfare*, edited by Daniel Marston and Carter Malkasian, 287-310. Oxford, UK: Osprey Publishing, 2010.
———. "The Role of Perceptions and Political reform in Counterinsurgency: The Case of Western Iraq, 2004-2005." *Small Wars and Insurgencies* 17, no. 3 (September 2006): 367-394.
Mansfield, Don. "The Irish Republican Army and Northern Ireland." In *Insurgency in the Modern World*, edited by O'Neill, Bard, 44-85. Boulder, CO: Westview Press, 1980.
Mao Tse-Tung. *On Guerrilla Warfare.* Translated by Samuel B. Griffith.

Chicago: University of Illinois Press, 1961.

———. *The Selected Military Writings of Mao Tse-Tung*. Peking: Foreign Language Press, 1972.

Markel, Wade. "Draining the Swamp: The British Strategy of Population Control." *Parameters* (Spring 2006): 35-48.

Marshall, S. L. A. "Thoughts on Vietnam." In *The Lessons of Vietnam*, edited by Thompson, W. Scott and Donaldson Frizzel, 46-55. New York: Crane, Russak and Company, 1977.

Marston, Daniel. "Adaptation in the Field: The British Army's Difficult Campaign in Iraq." *Security Challenges* 6, no. 1 (Autumn 2010): 71-84.

———. "The Indian Army, Partition, and the Punjab Boundary Force, 1945-1947." *War in History* 16, no.4 (2009): 469-505.

———. "Lost and Found in the Jungle." In *Big Wars and Small Wars*, edited by Strachan, Hew, 96-114. London: Routledge, 2006.

———. "Realizing the Extent of Our Errors and Forging the Road Ahead: Afghanistan 2001-2010." In *Counterinsurgency in Modern Warfare*, edited by Daniel Marston and Carter Malkasian, 251-286. Oxford: Osprey Publishing, 2010.

Marston, Daniel, and Carter Malkasian, eds. *Counterinsurgency in Modern Warfare*. Oxford, UK: Osprey Publishing, 2008.

McCoy, Alfred. *Policing America's Empire*. Madison, WI: University of Wisconsin Press, 2009.

McCuen, John J. *The Art of Counter-Revolutionary War*. Harrisburg, PA: Stackpole Books, 1966.

McGrath, John J. *Fire for Effect: Field Artillery and Close Air Support in the US Army.* Fort Leavenworth, KS: Combat Studies Institute Press, 2007.

McKenney, Janice E. *The Organizational History of Field Artillery 1775-2003.* Washington, DC: US Army Center for Military History, 2007.

Millen, Raymond. "Time for a Strategic and Intellectual Pause in Afghanistan." *Parameters* (Summer 2010): 33-45.

Metz, Steven. "New Challenges and Old Concepts: Understanding the 21st Century Insurgency." *Parameters* (Winter 2007-2008): 20-32.

Moghaddam, Fathali M. *The New Global Insecurity*. Santa Barbara, CA: Praeger Security International, 2010.

Moyar, Mark. *A Question of Command*. New Haven, CT: Yale University Press, 2009.

———. "Development in Afghanistan's Counterinsurgency." *Small Wars Journal*. www.smallwarsjournal.com/blog/2011/03/development-in-afghanistans-co/ (accessed 21 March 2011)

———. *Triumph Forsaken: The Vietnam War, 1954-1965.* New York: Cambridge University Press, 2006.

Nagl, John. "Counterinsurgency in Vietnam: American Organizational Culture and Learning." In *Counterinsurgency in Modern Warfare*, edited by Daniel Marston and Carter Malkasian, 119-136. Oxford, UK: Osprey, 2008.

———. *Learning to Eat Soup With a Knife: Counterinsurgency Lessons From Malaya and Vietnam.* Chicago: University of Chicago Press, 2002.

Nagl, John, and Paul L. Yingling. "The FA in the Long War: A New Mission in

COIN." *FA Journal* (July-August 2006): 33-36.

Najim Abed Al-Jabouri, and Sterling Jensen. "The Iraqi and AQI Roles in the Sunni Awakening." *Prism* 2, no. 1 (2010): 3-18.

Norland, Rod. "A Radical Cleric Gets Religion." *Newsweek*, 10 November 2007. www.newsweek.com/2007/11/10/a-radical-cleric-gets-religion.html (accessed 8 June 2011).

Omrani, Bijan. "The Durand Line: History and Problems of the Afghan-Pakistan Border." *Asian Affairs* 40, no. 2 (2009): 177-195.

O'Neill, Bard. "Revolutionary War in Oman." In *Insurgency in the Modern World*, edited by O'Neill, Bard, 213-234. Boulder, CO: Westview Press, 1980.

O'Neill, Mark. *Confronting the Hydra.* Sydney, AU: Lowy Institute, 2009.

Ott, David Ewing. *Vietnam Studies: Field Artillery, 1954-1973.* Washington, DC: Department of the Army, 1975.

———. "Vietnamization: FA Assistance Programs." *Field Artillery* (January-February 2007): 34-41.

Packer, George. "The Lesson of Tal Afar." *The New Yorker* 82, issue 8 (10 April, 2006). www.newyorker.com/archive/2006/04/10/060410fa_fact2 (accessed 13 May 2011).

Paget, Julian. *Counter-insurgency Operations: Techniques of Guerrilla Warfare.* New York: Walker and Company, 1967.

Paret, Peter, and John Shy. "Guerrilla Warfare and U.S. Military Policy: A Study." In *The Guerrilla and How to Fight Him: Selections from the Marine Corps Gazette*, edited by T. N. Green. New York: Praeger, 1962.

Paret, Peter. "Clausewitz," In *Makers of Modern Strategy*, edited by Peter Paret, 186-213. Princeton, NJ: Princeton University Press, 1986.

———. *French Revolutionary Warfare from Indochina to Algeria: The Analysis of a Political and Military Doctrine.* New York: Praeger, 1964.

———. *Makers of Modern Strategy.* Princeton, NJ: Princeton University Press, 1986.

Parker, Geoffrey. "The Western Way of War." In *The Cambridge History of Warfare,* edited by Geoffrey Parker, 1-11. New York: Cambridge University Press, 2005.

Porch, Douglas. Douglas Porch, "Bugeaud, Gallieni, Lyatuey: The Development of French Colonial Warfare." In *Makers of Modern Strategy*, edited by Peter Paret, 376-407. Princeton, NJ: Princeton University Press, 1986.

———. "French Imperial Warfare 1945-62." In *Counterinsurgency in Modern Warfare*, edited by Daniel Marston and Carter Malkasian, 79-100. Oxford, UK: Osprey Publishing, 2008.

Porter, Patrick. *Military Orientalism: Eastern War Through Western Eyes*. New York: Columbia University Press, 2009.

Price, D. L. *Oman: Insurgency and Development.* London: Institute for the Study of Conflict, 1975.

Pye, Lucian W. *Guerrilla Communism in Malaya*. Princeton, NJ: Princeton University Press, 1956.

Race, Jeffrey. *War Comes to Long An*. Berkeley, CA: University of California Press, 1972.

Ramsey, Robert. *Savage Wars of Peace: Case Studies of Pacification in the*

Philippines, 1900-1902. Fort Leavenworth, KS: Combat Studies Institute, 2007.

Ratnesar, Romesh, and Michael Weisskopf. "Person of The Year 2003: Portrait Of A Platoon." *Time*, 29 December 2003.

Raymond, William M., Keith R. Beurskens, and Steven M. Carmichael. "The Criticality of Captains' Education, Now and in the Future." *Military Review* (November-December 2010): 51-57.

Record, Jeffrey. *Bounding the Global War on Terrorism.* Carlisle, PA: Strategic Studies Institute, 2003.

Reiner, Rob. *This is Spiñal Tap*. Wilmington, NC: DiLaurentis/Embassy Pictures, 1984.

Ricks, Thomas E. *Fiasco*. New York: The Penguin Press, 2006.

———. *The Gamble*. New York: The Penguin Press, 2009.

Rubin, Barnett, and Ahmed Rashid. "The Great Game to the Great Bargain." *Foreign Affairs* 87, no. 6 (November-December 2008): 30-44.

Russell, James A. *Innovation, Transformation, and War: Counterinsurgency Operations in Anbar and Ninewah, Iraq, 2005-2007*. Stanford, CA: Stanford Security Studies, 2011.

Scales, Robert H. "Artillery's Failings in the Iraq War." *Armed Forces Journal* (November 2003): 44-48.

———. *Firepower in Limited War*. Novato, CA: Presidio Press, 1995.

Scott, Peter Dale. "*Vietnamization and the Drama of the Pentagon Papers.*" In *The Pentagon Papers Volume V*, edited by Noam Chomsky and Howard Zinn, 211-247. Boston: Beacon Press, 1972.

Selth, Andrew. "Ireland and insurgency: the lessons of history." *Small Wars & Insurgencies* 2, no. 2 (1991): 299-322.

Semple, Michael, and Fotini Christia. "How to Flip the Taliban." *Foreign Affairs* (July-August 2009).

Sepp, Kalev. "Best Practices in Counterinsurgency." *Military Review* (May-June 2005): 8-12.

Shy, John. "Jomini." In *Makers of Modern Strategy*, edited by Peter Paret, 143-185. Princeton, NJ: Princeton University Press, 1986.

Shy, John, and Thomas Collier. "Revolutionary war." In *Makers of Modern Strategy: Military Thought from Machiavelli to the Nuclear Age*, edited by Peter Paret, 815-862. Princeton, NJ: Princeton University Press, 1986.

Smith, E. D. *Counter-Insurgency Operations 1: Malaya and Borneo*. London: Ian Allan, 1985.

Stubbs, Richard. "From Search and Destroy to Hearts and Minds: The Evolution of British Strategy in Malaya 1948-60." In *Counterinsurgency in Modern Warfare*, edited by Daniel Marston and Carter Malkasian, 101-118. Oxford, UK: Osprey Publishing, 2010.

Sun Tzu. *Manual for War*. Translated by T. W. Kuo. Chicago: ATLI Press, 1989.

Thompson, Robert. *Defeating Communist Insurgency*. London: Chatto and Windus, 1967.

———. *No Exit From Vietnam*. New York: David McKay Company, 1969.

Thompson, W. Scott, and Donaldson Frizzell. *The Lessons of Vietnam*. New York: Crane, Russak and Company, 1977.

Thornton, Rod. "Getting It Wrong: The Crucial Mistakes Made in the Early

Stages of the British Army's Deployment to Northern Ireland." *Journal of Strategic Studies* 30, no. 1 (February 2007): 73-107.

Townshend, Charles. "The IRA and the Development of Guerrilla Warfare." *English Historical Review* 94 (April 1979): 318-345.

Trinquier, Roger. *Modern Warfare: A French View of Counterinsurgency.* Westport, CT: Praeger, 1964.

Tripodi, Christian. "Good for One But Not the Other; The Sandeman System of Pacification as Applied to Baluchistan and the North-West Frontier, 1877-1947." *The Journal of Military History* 73, no. 3 (July 2009): 767-802.

Tripp, Charles. *A History of Iraq.* New York: Cambridge University Press, 2005.

Tucker, Michael S., and Jason P. Conroy. "Maintaining the Combat Edge." *Military Review* 91, no. 3 (May - June 2011): 8-15.

US Army Artillery and Missile School. "Aerial Rocket Artillery." *Artillery Trends* 37 (January 1967): 41-44.

———. "Lessons Learned in Vietnam: 6400-mil Traverse for the 175mm Gun and 8-Inch Howitzer." *Artillery Trends* 37 (January 1967): 17-22.

———. "Riverine Artillery." *Artillery Trends* 39 (January 1968): 14-24.

US Army Fires Center of Excellence. *2011 Fire Seminar: State of Fires.* Symposium at Lawton, OK: Cameron University, 17 May 2011.

Ucko, David. *The New Counterinsurgency Era: Transforming the US Military for Modern Wars.* Washington, DC: Georgetown University Press, 2009.

Van Fleet, Townsend A. "105 in the Jungle." *Artillery Trends* 26 (October 1962): 19-26

Walter, Dennis I. "First Round Smoke." *Artillery Trends* 39 (January 1968): 25-27.

West, Bing. *No True Glory: A Frontline Account of the Battle for Fallujah.* New York: Bantam Books, 2005.

Willbanks, James. *Abandoning Vietnam.* Lawrence, KS: University of Kansas Press, 2004.

Wither, J. K. "Basra's Not Belfast." *Small Wars and Insurgencies* 20, no. 3 and 4 (September-December 2009): 622.

Wood, J. R. T. "Countering the Chimurenga: The Rhodesian Counterinsurgency Campaign." In *Counterinsurgency in Modern Warfare*, edited by Daniel Marston and Carter Malkasian, 191-208. Oxford, UK: Osprey Publishing, 2008.

Yates, Larry. *Field Artillery in Military Operations Other Than War: An Overview of the US Experience.* Fort Leavenworth, KS: Combat Studies Institute Press, 2005.

Zabecki, David T. *Steel Wind: Colonel Georg Bruchmuller and the Birth of Modern Artillery.* Westport, CT: Praeger, 1994.

www.ingramcontent.com/pod-product-compliance
Lightning Source LLC
Chambersburg PA
CBHW051123160426
43195CB00014B/2325